Advances in

Geosciences

Volume 10: Atmospheric Science (AS)

ADVANCES IN GEOSCIENCES

Editor-in-Chief: Wing-Huen Ip *(National Central University, Taiwan)*

A 5-Volume Set

Volume 1: Solid Earth (SE)
ISBN-10 981-256-985-5

Volume 2: Solar Terrestrial (ST)
ISBN-10 981-256-984-7

Volume 3: Planetary Science (PS)
ISBN-10 981-256-983-9

Volume 4: Hydrological Science (HS)
ISBN-10 981-256-982-0

Volume 5: Oceans and Atmospheres (OA)
ISBN-10 981-256-981-2

A 4-Volume Set

Volume 6: Hydrological Science (HS)
ISBN-13 978-981-270-985-1
ISBN-10 981-270-985-1

Volume 7: Planetary Science (PS)
ISBN-13 978-981-270-986-8
ISBN-10 981-270-986-X

Volume 8: Solar Terrestrial (ST)
ISBN-13 978-981-270-987-5
ISBN-10 981-270-987-8

Volume 9: Solid Earth (SE), Ocean Science (OS) & Atmospheric Science (AS)
ISBN-13 978-981-270-988-2
ISBN-10 981-270-988-6

A 6-Volume Set

Volume 10: Atmospheric Science (AS)
ISBN-13 978-981-283-611-3
ISBN-10 981-283-611-X

Volume 11: Hydrological Science (HS)
ISBN-13 978-981-283-613-7
ISBN-10 981-283-613-6

Volume 12: Ocean Science (OS)
ISBN-13 978-981-283-615-1
ISBN-10 981-283-615-2

Volume 13: Solid Earth (SE)
ISBN-13 978-981-283-617-5
ISBN-10 981-283-617-9

Volume 14: Solar Terrestrial (ST)
ISBN-13 978-981-283-619-9
ISBN-10 981-283-619-5

Volume 15: Planetary Science (PS)
ISBN-13 978-981-283-621-2
ISBN-10 981-283-621-7

Advances in

Geosciences

Volume 10: Atmospheric Science (AS)

Editor-in-Chief

Wing-Huen Ip

National Central University, Taiwan

Volume Editor-in-Chief

Jai Ho Oh

Pukyong National University, Korea

Gyan Prakash Singh

Banaras Hindu University, India

World Scientific

NEW JERSEY • LONDON • SINGAPORE • BEIJING • SHANGHAI • HONG KONG • TAIPEI • CHENNAI

Published by

World Scientific Publishing Co. Pte. Ltd.

5 Toh Tuck Link, Singapore 596224

USA office: 27 Warren Street, Suite 401-402, Hackensack, NJ 07601

UK office: 57 Shelton Street, Covent Garden, London WC2H 9HE

British Library Cataloguing-in-Publication Data
A catalogue record for this book is available from the British Library.

ADVANCES IN GEOSCIENCES
A 6-Volume Set
Volume 10: Atmospheric Science (AS)

ISBN-13 978-981-283-610-6 (Set)
ISBN-10 981-283-610-1 (Set)
ISBN-13 978-981-283-611-3 (Vol. 10)
ISBN-10 981-283-611-X (Vol. 10)

Typeset by Stallion Press
Email: enquiries@stallionpress.com

Printed by FuIsland Offset Printing (S) Pte Ltd. Singapore

EDITORS

REVIEWERS

The Editors of Volume 10 would like to acknowledge the following referees who had helped review the papers published in this volume:

Prof. Dongsong Sun
Dr. V. Rao Kotamarthi
Dr. B. K. Spara
Prof. Choo Hie Lee
Dr. (Mrs.) Ashwani Kulkarni
Prof. Jun Mustsumoto
Prof. Qian Yongfu
Dr. X. Li-Jones
Dr. Tomoaki Nishizawa
Prof. Hu Hanling
Dr. Jiangyu Mao
Dr. Kripalani
Prof. P. N. Sen
Dr. G. P. Singh
Prof. Bimblecombe
Dr. Zhou Tianjun

CONTENTS

Advances in Geosciences
Vol. 10: Atmospheric Science (2007)
Eds. J. H. Oh and G. P. Singh

SEASONAL VARIATION IN THE STRUCTURE OF QTP ATMOSPHERIC HEAT SOURCE IN 1961–2001

ZHONG SHANSHAN, HE JINHAI, GUAN ZHAOYONG
and LIU XUANFEI
Key Laboratory of Meteorological Disasters of Jiangsu,
Nanjing University of Information Science and Technology,
Nanjing, China 210044

ECMWF daily reanalysis is applied to investigate 1961–2001 heat source/sink and the climate features in relation to the atmospheric heat distribution over the QTP (Qinghai–Tibetan Plateau) by means of the "inverse algorithm." Results suggest that (1) in March–September (October–February), the QTP acts as a heat (cold) source, the strongest being in June (December). The heating effects of the QTP are asymmetric in the seasons; (2) as shown in the heating vertical profile, the maximum heat source layer occurs dominantly between 500 and 600 hPa, but with the season-dependent heating strength and depth, and, in contrast, the cold source has its maximum layer and intensity varying as a function of time; (3) the horizontal distribution of the heat sources throughout the troposphere $\langle Q_1 \rangle$ (from surface to 100 hPa) is complicated, displaying noticeable regionality; (4) since 1979 the seasonal variability of the heat source has shown climate transition signals, as clearly seen in 1990/1991.

1. Introduction

The QTP is a region higher in elevation and more complicated in surface feature than any other area in the world, called a "third pole" next to the Arctic and Antarctic in climate. The tremendous prominence of the QTP has changed the pattern of Asian climate, exerting pronounced effects on atmospheric circulations and climate worldwide.[1]

Abnormal change in the QTP thermal condition in the atmosphere bears a close relation to East-Asian circulations,[2] acting as an indicator of summer precipitation over the Jiang-Huai valley, South China and North China.[3,4] When the QTP heat source is strengthened, the rainfall increases in the upper Yangtze and Huaihe valleys as opposed to Southeast and North China. Around seasonal transition, the differences in the source/sink over the QTP with its vicinity are one of the causes of heavy flood/drought happening in the Jiang-Huai valley.[5,6]

As regards the effects of the QTP thermal role and its ensuing moist process upon monsoon features, numerous Chinese researchers have demonstrated[7,8] that the variation in nonadiabatic heating in the transition early in summer gives rise to the change in land–sea thermal contrast, thus providing a favorable background for Asian summer monsoon onset, which has significant impacts on the outbreak.[9] As shown by Jiang and Luo,[10] when the East Asian monsoon begins, the nonadiabatic heating is responsible for tropospheric explosive warming over the Southeast QTP, leading to change in temperature gradient southward of the east QTP, consequently resulting in the adjustment of wind field for the onset of the east Asian monsoon. Jian[11] indicated that the May–June conspicuous warming due to nonadiabatic heating in the mid-higher troposphere over the east QTP is of much importance to the northward march and maintenance of the summer monsoon. Differences in space/time distribution of QTP summer rainfall also cause variations in the spatial/temporal distribution of the heat sources over the QTP and its neighborhood — the variations make atmospheric circulation change accordingly, finally leading to the difference in the onset time of the monsoon.[12] Although monsoon researchers, domestic and foreign, diverge about how the QTP sensible heating affects the summer monsoon onset, undoubtedly, the QTP heating represents one of the mechanisms of the monsoon onset.

For lack of observations, how to obtain correct calculation is the linchpin of studies. The 1961–2001 ECMWF (ERA hereafter) daily reanalysis was employed to calculate the heat source features. The atmospheric apparent heat source Q_1 was found by the use of the "inverse algorithm" developed by Yanai.[13] Q_1 comprised three terms: local term, advection term, and vertical transport term.

2. Regional Mean Climate Condition Due to the QTP Heat Sources

To gain insight into the tropospheric thermal regime 1961–2001 $\langle Q_1\rangle$-associated mean climate condition was analyzed. Over 1961–2001, within the region of 3000 m, 41-yr monthly climate mean conditions of $\langle Q_1\rangle$ were shown by full line with open circles (Fig. 1), indicating that in March–September, $\langle Q_1\rangle > 0$ as the heat source began intensification from March, maximizing at 214 W/m^2 in June, and decreasing thereafter; in October–February, $\langle Q_1\rangle < 0$ suggestive of a cold source, the strongest being in

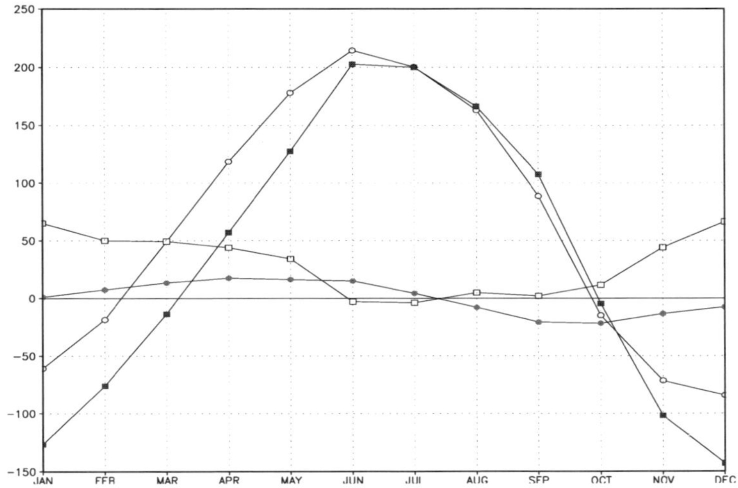

Fig. 1. Regional averaged monthly $\langle Q_1 \rangle$ (W/m^2) over the QTP in 1961–2001 (the solid line with open circles on for $\langle Q_1 \rangle$, the full line with solid circles upon for the local term, full line with open squares on for the advection term, and the full line with solid squares upon for the vertical transport term).

December (about $-84\,\text{W/m}^2$). It follows that for $\langle Q_1 \rangle$, the yearly thermal regime displayed a longer period of it as a heat source (on the order of 7 months), with the value much higher (250% as strong in absolute value) compared to the cold source (214 vs $-84\,\text{W/m}^2$), exhibiting asymmetry of the annual cycle.

Comparing the three terms of $\langle Q_1 \rangle$, we found that the vertical transport term makes the greatest contribution to $\langle Q_1 \rangle$.

The QTP is responsible for noticeably lifting the heat source in the troposphere because of its great elevation. What does the height-varying heating profile look like over the Plateau? Figure 2 presents the height-evolving monthly mean Q_1 and its components in 1961–2001 over the QTP at an elevation of 3000 m.

The height-varying Q_1 is featured mainly by the opposite trend of intensity of the heat to cold source, and the whole process can be described as a "cylinder stator" of an engine in operation, with the piston representing source transition that divides the heat and cold source in vertical, as shown in the "steam chests" of the engine. As time goes on, the volume (thickness) is changing constantly for both. As the heat source expands its volume, i.e. the piston goes up, the thickness of the cold source diminishes and

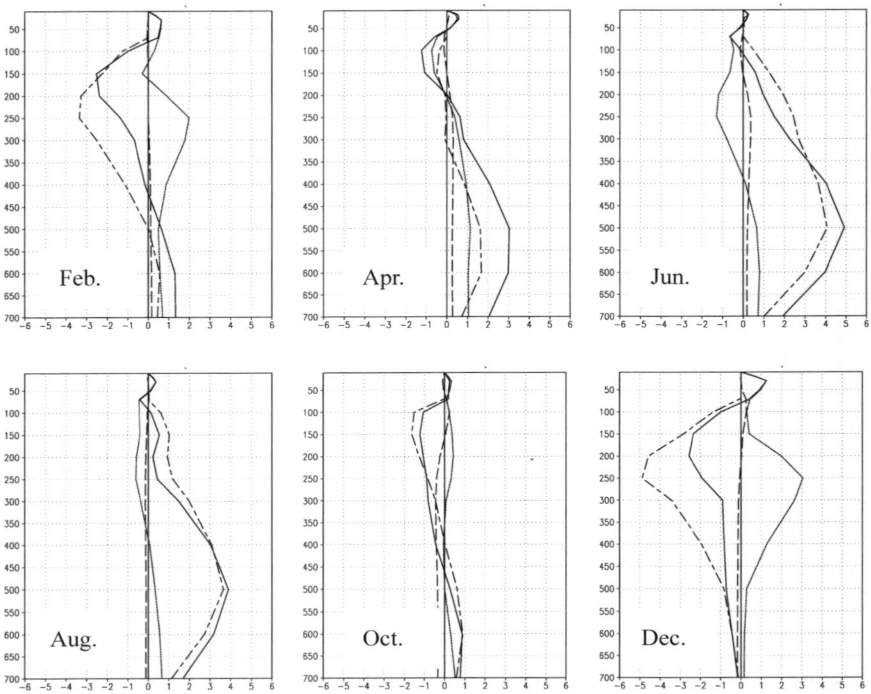

Fig. 2. The vertical profile of monthly mean Q_1 and its components at 3000 m level over the QTP, with solid line for Q_1, long dashed for the local term, dotted line for the advection term, and dash dotted for the vertical transport term (Units: K day^{-1}).

vice versa. As indicated by the vertical profile, the "piston" reaches its top in July–August when the heat source is the deepest. Conversely, as the "piston" has its lowest position, the cold source covers the greatest depth in October–December, and the troposphere is nearly under the control of the cold source.

The strongest heat source layer occurs almost at 500–600 hPa, except for its intensity that peaks in June and decreases toward January and December (as shown in Fig. 3), with the cold source dominating practically all atmospheric levels in December. Note that the height and vigor of the maximal cold source layer change with time. In June–August, the cold source layer is weak in intensity and shallow in depth, reaching its greatest thickness in cold months (NDJ), with its highest strength at 200 hPa.

The vertical profiles of monthly Q_1 and its components (Fig. 2) show that the height-dependent local term value is considerably smaller

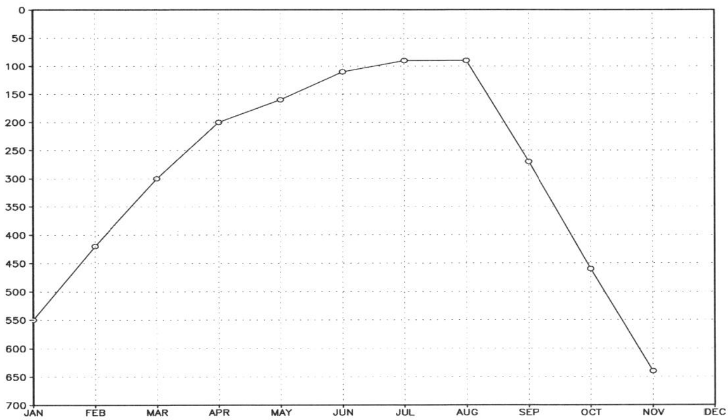

Fig. 3. The height of heat source layer varies with season over the QTP.

compared to the other two, and so we ignored it. Consequently, the terms affecting Q_1 are the terms of advection and vertical transport. The summer Q_1 profile is similar to that of the vertical term that makes the dominant contribution thereto in summer. In winter the atmospheric cold source is produced as a result of the terms of advection and vertical transport, opposite in phase, with the latter marginally larger.

3. The Horizontal Distribution of $\langle Q_1 \rangle$

The heat source distributed throughout the extent $\langle Q_1 \rangle$ over the QTP is given in Fig. 4. In January (figure omitted), the cold source covers almost the entire region at 3000 m level of the QTP, with the core in the Southeast QTP. In February and March, the QTP cold center remains constant, with its domain contracted southward. In April, the whole QTP is under the effect of a heat source, centered on 85°E and 30°–35°N, whose intensity reaches >150 W/m^2. The former cold source center has been covered with a heat source, with its value lower compared to the surroundings. With 90°E as the division, we see that the eastern intensity is lower than the western one. In May (figure omitted), the QTP heat source continues to intensify, with its center remaining at the western QTP west of 90°E and the 200 W/m^2 core stretches Southeast. In June, the eastern heat source experiences sudden reinforcement, arriving at >200 W/m^2 as its central value except that the 200 W/m^2 isoline takes a more northern position in the western than in the eastern QTP, suggesting that the western is stronger

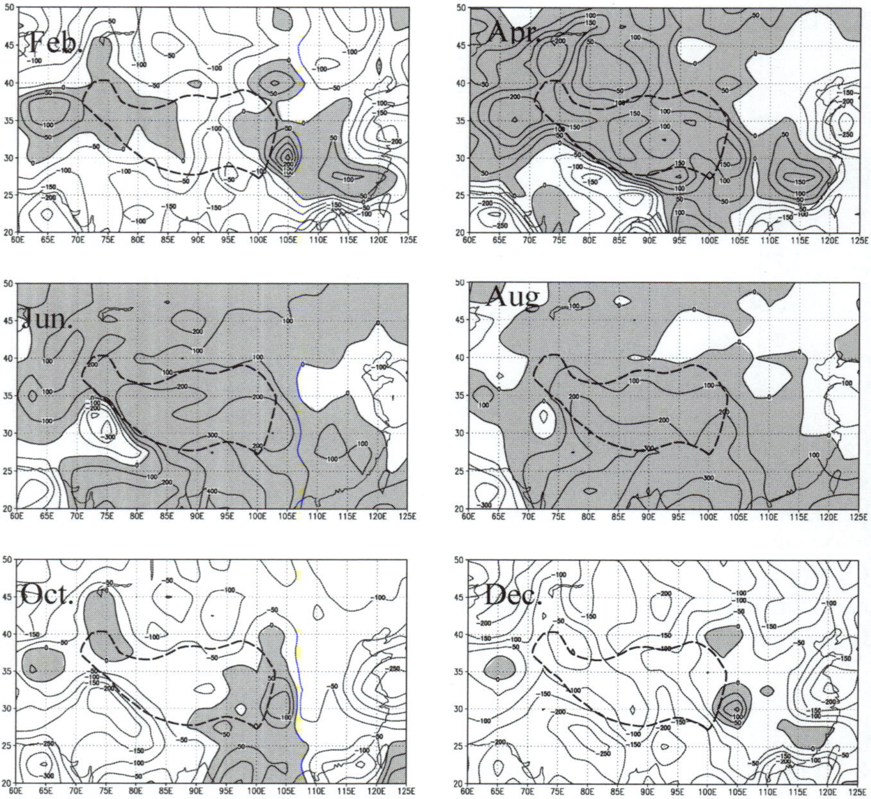

Fig. 4. Monthly mean $\langle Q_1 \rangle$ distribution in horizontal over the QTP (Units: W/m^2).

than the eastern heat source. At this time, the QTP heat source is being the strongest in the months. In July (figure omitted), concurrently with the weakening of the source to the north of the Bay, the QTP $200\,W/m^2$ isoline begins a southward withdrawal. As August arrives, concurrently with the southward retreat of the $200\,W/m^2$ contour, a break occurs to the heat source at the border between the Himalayas and the North and South (Hengduan) ranges. In September (figure omitted), in pace with the southward contraction of the Indo-China heat source, the counterpart of the western QTP begins to weaken swiftly, leading to the $100\,W/m^2$ contour Northeast–Southwest directed, implying the eastern heat source stronger than the western one of the QTP, with the cores located, respectively, in the western Sichuan Basin and around the eastern Himalayas. As October arrives, there appears a cold source in the west and a weak heat source in the east of the QTP. As November (figure omitted) sets in, the QTP is under the control of cold sources distributed in a similar pattern to that in January. In December, the cold source is the strongest, centered at the southeast QTP, with its values lower than $-150\,\mathrm{W/m^2}$.

To sum up, $\langle Q_1 \rangle$ horizontal distribution shows that from April to August, the heat source is stronger in the west than in the east, with the contour located more northward in the west than in the east. In Spring, the western heat source intensifies rapidly and not until May–June does the eastern one do so. The center of $200\,\mathrm{W/m^2}$ appears in May in the west, but in June in the east. Starting from July, the heat sources begin weakening southward, with the western one abating quickly. The heat source changes into a cold one in the west (east) in October (November).

4. The $\langle Q_1 \rangle$ Space/Time Variations

Only the 1979–2001 data were used for EOF study. The $\langle Q_1 \rangle$ was separated into four seasons, and the mode of greater variance contribution (EOF1) was taken out for discussion, as shown in Fig. 5.

In Spring, the $\langle Q_1 \rangle$ spatial pattern of EOF1 displays anti-phase change in the central–Northeast and Southeast QTP, the variability center value in the central source being twice as high as in the Southeast counterpart. From the time series, we see that a positive (negative) amplitude occurs in

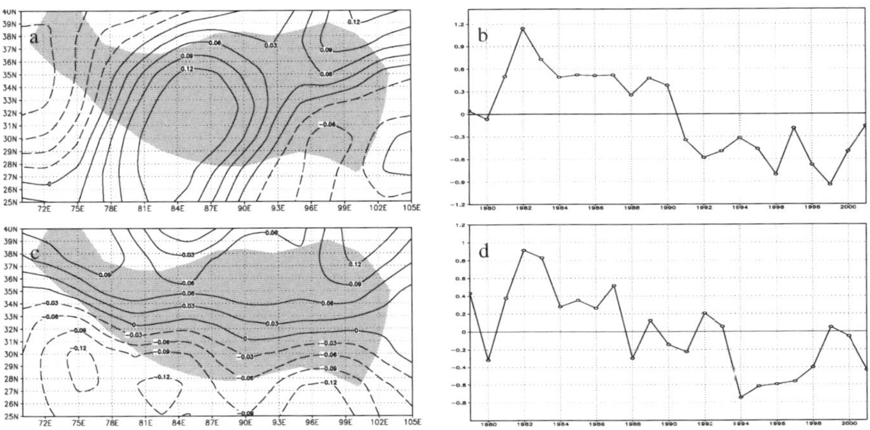

Fig. 5. EOF analysis of the QTP $\langle Q_1 \rangle$, giving the EOF1 Spring space pattern in (a), the timeseries in (b), the Summer pattern in (c), and the timeseries in (d).

1981–1990 (1991–2001), suggesting that the trend of anomalous change in the heat source occurs over 1990/1991. Prior to 1990 the central–Northeast QTP is inside the positive anomaly variability center as opposed to the situation after the year, with an opposite change in the Southeast source.

In summer, $\langle Q_1 \rangle$ EOF1 shows a see-saw form divided by 32.5°N, positive in the north and negative values in the south. The maximal variability center was not over the QTP but to the south, where there were three cores. The temporal sequence also indicates the climate change of the anomalous variation in the heat source, with positive (negative) values in 1981–1987 (1994–1998), with 1988–1993 as the transition stage, i.e. south of 32°N there occurred negative anomaly of variability in summer in the 1980s and positive anomaly was from the end of the 1980s to early 1990s.

The EOF1 Winter/Autumn space patterns are similar to that of Spring (figure not shown), showing the anti-phase distribution between the central and Northeast and Southeast QTP, with the 1990/1991 as the climate transition year.

It follows that the EOF1 — given seasonal $\langle Q_1 \rangle$ indicates the climate transition, differing in that the variability center is kept over the QTP in all but summer seasons and it is on the south side of the QTP in Summer, i.e. over the Indo-China and northern Indian Peninsula. This may be due to the fact that the summer QTP heat source is not an independent center but part of the source in the Indo-China and Indian Peninsula as well as the Bay of Bengal in between.

5. Conclusions

The study calculated heat source and heat sink over the Tibetan Plateau and its vicinity (QTP) during 1961–2001 using the ERA daily reanalysis and the "inverse algorithm," and discussed the climate regimes linked to the thermal source over the QTP.

Some reasonable conclusions are obtained, that is, the region over the QTP with the height more than 3000 m above the sea level acts as a heat source, and as a heat sink during October–February. The heat source lasts for 7 months in the whole atmospheric extent and much stronger than the sink in the wintertime. Therefore, the heating effects of the QTP are asymmetric in the seasons. We also discussed the space extent in the vertical and time variation of the heat source and sink over the QTP. It is found that during April–August, the heat source is stronger in the west than in the east. As Spring arrives, the western heat source increases rapidly while the eastern one begins to rapidly increase during May–June, with the 200 W/m^2 core value appearing in May for the west and in June for the east. The QTP heat source starts to decrease and withdraw southward from July. The western heat source abates faster, changing into a heat sink in October, 1 month earlier compared to the eastern one. It is depicted that since 1979 the seasonal variability of the heat source has shown climate transition signals, as clearly seen in 1990/1991.

Acknowledgments

The work was supported by a key item of National Natural Science Foundation of China (40633018) and graduate student innovation planning project in Jiangsu of China in 2006.

References

1. G. X. Wu, Y. M. Liu, X. Liu, A. M. Duan and X. Y. Liang, How the heating over the Tibetan Plateau affects the Asian climate in summer, *Chin. J. Atmos. Sci.* **29**(1) (2005) 47–56.
2. A. M. Duan, Y. M. Liu and G. X. Wu, Heat condition over Qinghai–Tibet Plateau in Apr.–Jun. and its effect on east Asia precipitation in mid-summer and abnormal atmospheric circulation, *Chin. Sci. (D)* **33**(10) (2003) 997–1004.
3. P. Zhao and L. X. Chen, Climate characteristics of Qinghai–Tibet Plateau atmospheric heat source in 35 years and its relation to Chinese precipitation, *Chin. Sci. (D)* **31**(4) (2001) 327–332.
4. S. R. Zhao, Z. S. Song and L. R. Ji, Heating effect of the Tibetan Plateau on rainfall anomalies over North China during rainy season, *Chin. J. Atmos. Sci.* **5** (2003) 881–893.
5. Y. Q. Lü and Y. F. Gong, Atmospheric heat source/sink change characteristics over Qinghai–Xizang Plateau and its vicinity region in summer of 2001 and 2003, *Plateau Meteorol.* **25**(2) (2006) 195–202.
6. Y. Zhao and Y. Qian, Relationships between the surface thermal anomalies in the Tibetan Plateau and the rainfall in the Jianghuai area in summer, *Chin. J. Atmos. Sci.* **31**(1) (2007) 145–154.
7. Q. G. Zhu and J. L. Hu, Numerical experiments on the influences of the Qinghai–Xizang Plateau topography on the summer general circulation and the Asian summer monsoon, *J. Nanjing Instit. Meteorol.* **16** (1993) 120–129.
8. J. H. He, J. Li and Q. G. Zhu, Sensitivity experiments on summer monsoon circulation cell in East Asia, *Adv. Atmos. Sci.* **6** (1989) 120–132.
9. X. Liu, G. X. Wu, Y. M. Liu and P. Liu, Diabatic heating over the Tibetan Plateau and the seasonal variations of the Asian circulation and summer monsoon onset, *Chin. J. Atmos. Sci.* **26**(6) (2002) 781–793.

10. N. B. Jiang and H. B. Luo, Effects of the heating of the Tibetan Plateau on the onset of the east Asian summer monsoon, *Acta Scientiarum Naturalium Universitatis Sunyatseni* **35**(suppl.) (1996) 194–199.
11. M. Q Jian and H. B. Luo, Heat sources over Qinghai–Xizang Plateau and surrounding areas and their relationships to onset of SCS summer monsoon in 1998, *Plateau Meteorol.* **20**(4) (2001) 381–387.
12. Y. F. Gong, L. R. Ji and T. Y. Duan, Precipitation character of rainy season of Qinghai–Xizang Plateau and onset over east Asia monsoon, *Plateau Meteorol.* **23** (2004) 313–322.
13. M Yanai, C. Li and Z. S. Song, Seasonal heating of the Tibetan plateau and its effects on the evolution of the Asian summer monsoon, *J. Meteorol. Soc. Jpn.* **70**(1) (1992) 319–350.

Advances in Geosciences
Vol. 10: Atmospheric Science (2007)
Eds. J. H. Oh and G. P. Singh

RAINFALL OVER THAILAND DURING ENSO (1997–2000)

WONLEE NOUNMUSIG* and PRUNGCHAN WONGWISES
The Joint Graduate School of Energy and Environment,
King Mongkut's University of Technology Thonburi,
126 Pracha-U-Thit Rd., Bangmod, Tungkru,
Bangkok 10140, Thailand
**wonlee_n@yahoo.com*

In this chapter, the yearly mean rainfall taken from the Thai Meteorological Department during 1972–2001 in each region of Thailand was analyzed comparing with 30 years' averaged rainfall in order to study the influence of ENSO. The rainfall in Thailand during ENSO 1997–2000 was selected as the case study. The results show that the amount of rainfall in most regions of Thailand during 1997/1998 (El Nino) is less than the long-term mean, while the amount of rainfall is strongly more than long-term mean for the whole Thailand in La Nina 1999/2000. The amount of rainfall during 1997–2001 shows the strong anomalies in early rainy season (May–June). Moreover, the Regional Atmospheric Model System (RAMS) version 6.0 is used to simulate the rainfall pattern in May and June 1997 and 1999. The trends of model results agree well with the observed data analysis.

1. Introduction

Thailand is situated in the southwestern part of Indo-Chinese Peninsula between latitudes 5° 37′N to 20° 27′N and longitudes 97° 22′E to 105° 37′E. The climate of Thailand is influenced by the southwest monsoon and northeast monsoon, which can be classified generally into three seasons: mid-October to mid-February of the next year is the moderate winter season, mid-February to mid-May is the summer season, and mid-May to mid-October is the rainy season. Her national economies are mainly the agricultural products, which largely depend on the weather and climate conditions. Favored by the southwest monsoon, plenty of rainfall is precipitated all over the tropical country of Thailand during the rainy season. This is the normal case. But in some abnormal years, stronger or weaker southwest monsoon may cause flood or droughty disaster, which affect the agricultural products.

The variation in rainfall is caused by many factors. One of the interesting factors is the ENSO phenomenon. Although the center of this event is in the equatorial Pacific, many researchers have indicated that the impacts of this phenomenon on the climates cover more than 75% of the earth.[1,4,5,8] Some of the examples are the change in the pattern of floods, droughts, cyclone, and severe storm activity.[3,6,13] There are two phases of ENSO: warm events (known as El Nino) and cold events (known as La Nina). In general, when El Nino event occurs, the first visible impact is an increase in rainfall in the eastern Pacific, including parts of South America, and a decrease in rainfall in the western Pacific locations such as Australia, Indonesia, Southeast Asia, and the Philippines. The amount of rainfall opposite to El Nino event, is on the La Nina event.[2,7] That is, a decrease in rainfall in the eastern Pacific and parts of South America, and an increase in rainfall in the western Pacific.

Recently, the regional climate model is increasingly being used for simulating the characteristic of rainfall. In this study, the case studied in ENSO during 1997–2000 was chosen to investigate the characteristic of rainfall in each region of Thailand. There are two events of ENSO during this period: El Nino year 1997–1998 and La Nina year 1999–2000. Basically, the area average is used to summarize the rainfall during the rainy season of Thailand. The anomaly months were simulated by the RAMS model.

2. Methodology

2.1. *The observed data analysis*

The monthly rainfall data during 1 January 1972–31 December 2001 are taken from Thai Meteorological Department. These data were selected to study the amount of rainfall over Thailand during the ENSO year and the normal year. All the observed stations were scattered in six regions including 16 stations in northern, 16 stations in northeastern, 10 stations in central, 10 stations in eastern, 14 stations in southern (east coast), and 5 stations in southern (west coast) part of Thailand.

2.2. *Model simulation*

The regional climate model based on the RAMS[9] version 6.0, developed at Colorado State University, was employed for these experiments. The basic equations were a set of nonhydrostatic compressive dynamic equations and a thermodynamic equation. A Kain–Fritsch cumulus parameterization scheme is used to simulate convective rainfall. Other parameterizations are standard for a simulation of this type. Surface fluxes of heat and moisture are represented through the Land Ecosystem Feedback land surface model.[14] The model domain has horizontal dimensions of 82×82 grid points with a grid spacing of 60 km, encompassing the Thailand and some part of Indian Ocean. The model uses a vertically stretched grid with a maximum vertical grid spacing of 1000 m. The minimum vertical grid spacing is 100 m with a vertical stretch ratio of 1.2. There are 30 grid points in the vertical one. The polar stereo-projection is used, and the center of the domain is located at 13.5N, 100E (Fig. 1). The meteorological initial and boundary conditions were interpolated from the National Centers for Environmental Prediction (NCEP) daily reanalysis (available online at http://www.cdc.noaa.gov/Datasets/ncep.reanalysis/pressure/).

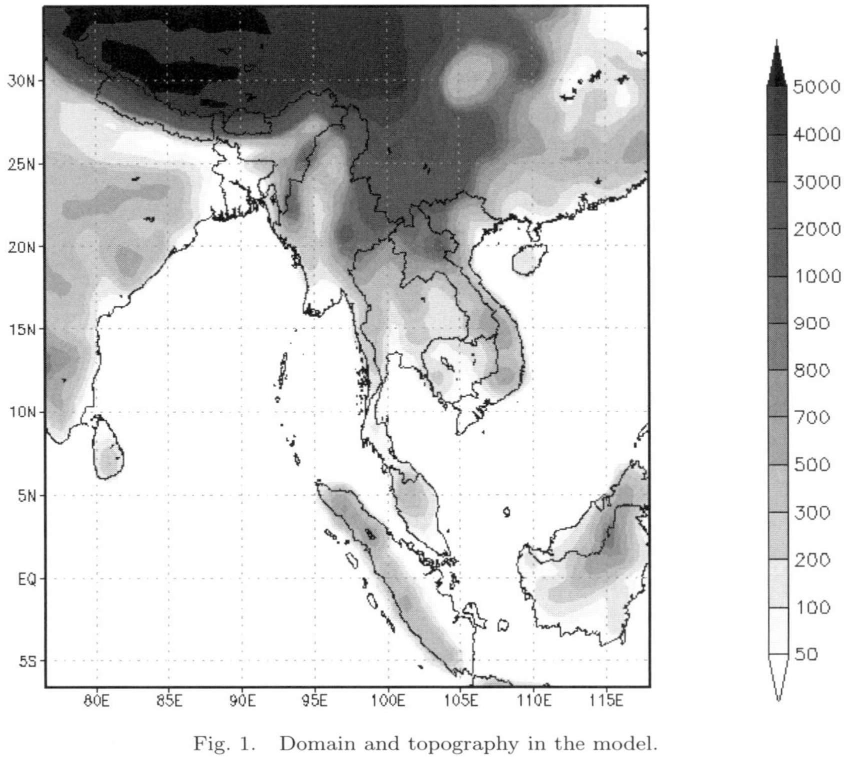

Fig. 1. Domain and topography in the model.

3. Results and Discussion

3.1. *Observed monthly rainfall during rainy season*

The ENSO during 1997–2000 were considered. During 1997–1998 El Nino started in March 1997 when the highest sea surface temperature occurred in September 1997, and the El Nino event ended in June 1998. Later, La Nina occurred in June 1998 and persisted through 2000.[4] The abnormal rainfall during the rainy season during 1997–2000 was calculated from the difference in the monthly rainfall with long-term mean as shown in Fig. 2. It can be seen that the abnormal rainfall during ENSO year 1997–2000 is strong during the early and end of the rainy season. During this period, the monthly rainfall during El Nino (1997) shows negative value (below normal), while positive value (above normal) was found in La Nina (1999–2000) in most of the regions in Thailand. The decrease or increase in rainfall during this period was caused by the movement of ITCZ, and is strong in monsoon. During El Nino (1997), it was found that the ITCZ moves rapidly within a short time in Thailand and becomes weak in monsoon, whereas in La Nina year (1999–2000) the ITCZ moves early to Thailand during the onset of early monsoon and reaches the peak in monsoon.[10–12]

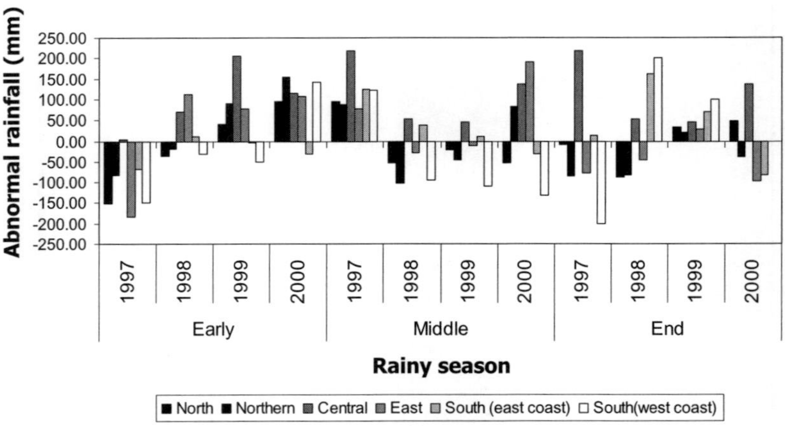

Fig. 2. The rainfall anomalies (mm) during rainy season 1997–2000.

3.2. *Model simulation*

The rainfall in each rain gauge stations was divided in each region of Thailand by using area-average. The choice of gauges was by selected stations of TMD which had complete data in May and June, 1997 and 1999. The time series of area-averaged daily total rainfall in six regions are compared with the observed value. The comparison is shown in Figs. 3(a)–3(f) for Northern, Northeastern, Central, Eastern, east coast

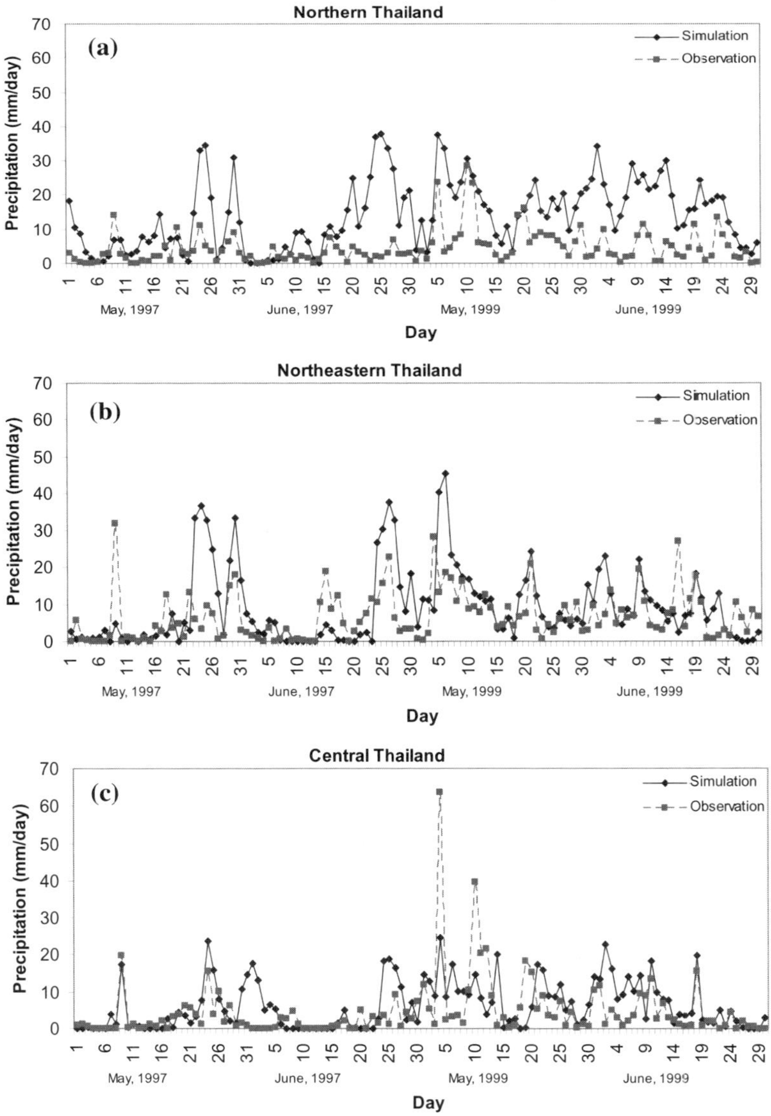

Fig. 3. The observed (dashed) and simulated (solid) area-averaged daily rainfall in 4 months for the six regions of Thailand.

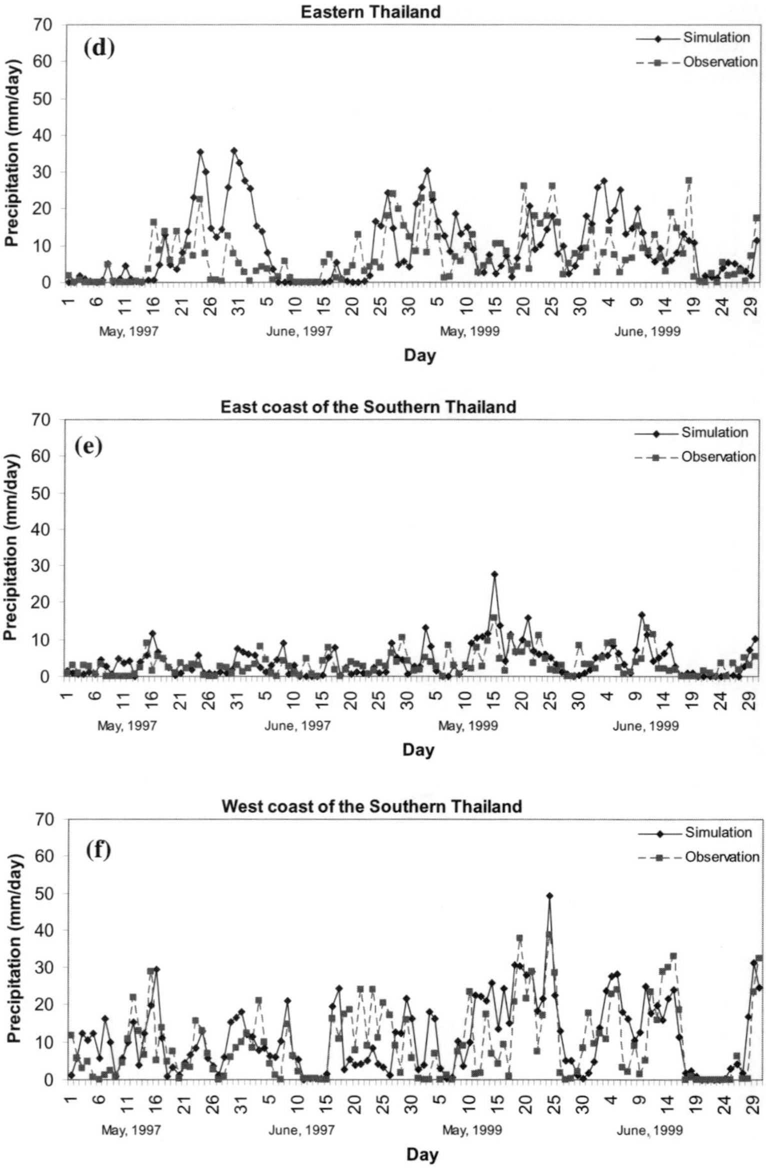

Fig. 3. (*Continued*).

of Southern, and west coast of Southern Thailand, respectively. All the plots show that the simulated rainfall amounts in each region were always higher than the observed value. However, the trend of simulated rainfall agrees with the observed rainfall. Both simulated and observed rainfall of all regions of Thailand show that the amount of rainfall in La Nina 1999 during these 2 months is more than that in El Nino 1997.

4. Conclusion

The total rainfall in Thailand during 1997–1998 (El Nino) was less than normal, while in La Nina year 1999–2000, the total rainfall was more than normal. During El Nino 1997–1998, it can be seen that the anomaly of rainfall occurred during the early rainy season, while during La Nina 1999–2000, the abundance of rainfall started in April and the amount of rainfall was more than the long-term mean for the whole of Thailand except the Southern (east coast) part of Thailand. By using RAMS to simulate the rainfall in May and June in El Nino 1997 and La Nina 1999, the simulated monthly rainfall was overestimated, but the trend of simulated daily rainfall in the model agrees with the observational data. It indicates that the amount of rainfall in La Nina 1999 during these 2 months is higher than El Nino 1997.

Acknowledgments

This work was supported by the Joint Graduate School of Energy and Environment, King Mongkut's University of Technology Thonburi. Special thanks are also due to Thai Meteorological Department (TMD) and the European Center for Medium Range Weather Forecasting (ECMWF) for the data supported.

References

1. R. J. Allan, ENSO and climatic variability in the past 150 years, in *El Nino and the Southern Oscillation Multiscale Variability and Global and Regional Impacts*, eds. H. F. Diaz and V. Markgraf (Cambridge University Press, Cambridge, 2000), pp. 3–55.
2. M. H. I. Dore, Climate change and changes in global rainfall patterns: What do we know? *Environ. Int.* **31** (2005) 1167–1181.

3. J. R. E. Harger, ENSO variations and drought occurrence in Indonesia and the Philippines, *Atmos. Environ.* **29** (1995) 1943–1955.
4. International Panel of Climate Change (IPCC), *Climate Change 2001 — Impacts, Adaptation, and Vulnerability* (Cambridge University Press, Cambridge, 2001).
5. P. D. Jones, The influence of ENSO on global temperatures, *Climate Monitor* (1988) 80–89.
6. R. H. Kripalani and A. Kulkarni, Climatic impact of El Nino/La Nina on the Indian monsoon: A new perspective, *Weather* **52** (1997) 39–46.
7. K. K. Kumar, B. Rajagopalan and M. A. Cane, On the weakening relationship between the Indian monsoon and ENSO, *Science* **284** (1999) 2156–2159.
8. Z. X. Long and C. Y. Li, GCM modeling of the impacts of the ENSO on east Asian Monsoon activities, *Acta Meteorologica Sinica* **57**(6) (1999) 663–671.
9. R. A. Pielke *et al.*, A comprehensive meteorological modeling system — RAMS, *Meteor. Atmos. Phys.* **49** (1992) 69–91.
10. TMD, *Rainy Season of Thailand for 1998* (Thai Meteoroogical Department, Thailand, 1998).
11. TMD, *Rainy Season of Thailand for 1999* (Thai Meteoroogical Department, Thailand, 2000).
12. TMD, *Rainy Season of Thailand For 2000* (Thai Meteoroogical Department, Thailand, 2001).
13. M. C. Wu, W. L. Chang and W. M. Leung, Impacts of El Nino; southern oscillation events on tropical cyclone landfalling activity in the Western North Pacific, *J. Climate* **17** (2004) 1419–1428.
14. R. L. Walko *et al.*, Coupled atmosphere — biophysics–hydrology models for environmental modeling, *J. Appl. Meteorol.* **39** (2000) 931–944.

Advances in Geosciences
Vol. 10: Atmospheric Science (2007)
Eds. J. H. Oh and G. P. Singh

TEMPORAL AND SPATIAL VARIATION OF CLOUD MEASURED WITH A PORTABLE AUTOMATED LIDAR

TATSUO SHIINA and TOSHIO HONDA
Faculty of Engineering, Chiba University,
1-33 Yayoi-cho, Inage-ku, Chiba-shi 263-8522, Japan

NOBUO TAKEUCHI, GERRY BAGTASA and HIROAKI KUZE
Center for Environmental Remote Sensing, Chiba University,
1-33 Yayoi-cho, Inage-ku, Chiba-shi 263-8522, Japan

AKIHIRO SONE and HIROFUMI KAN
Hamamatsu Photonics, 5000 Hirakuchi, Hamakita-ku
Hamamatu-shi, Shizuoka-ken, 434-8601, Japan

SUEKAZU NAITO
Chiba Prefecture Environmental Research Center,
1-8-8 Iwasakinishi, Ichikawa-shi 290-0046, Japan

A portable automated lidar (PAL) system, which conducts full-time operation and all-weather observation through the laboratory window, has been developed. Observations of long-term temporal and spatial dynamics of the atmosphere are described and the advantage of full-time operation is discussed.

1. Introduction

Atmospheric convection has an effect on cloud formation, and it leads to heavy rain or lighting strike. It also affects diffusion of suspended substances. Changes in the atmospheric convection due to the climate change may influence the large- and local-scale transformation of particles such as the yellow sand. In this context, it is essential to understand the temporal and spatial dynamics of the atmosphere, which cannot be monitored with conventional, fixed-point observation systems or meteorological satellites.

Lidar is an appropriate tool for monitoring time and spatial dynamics of the atmosphere, especially aerosols and clouds. Although various kinds of lidar systems have so far been developed, observations are limited in terms of time spans. Besides, observation directions are usually fixed both horizontally and vertically. These limitations originated from the system stability as well as the complication of system maintenance including the laser device.

A micro-pulse lidar (MPL), developed by Spinhirne in 1993, is a compact lidar system that provides easy operation and long-term observation.[1] Using a laser-diode pumped laser of micro-joule output energy, MPL ensures the eye-safety features. Signal-to-noise ratio was improved by narrowing the receiver field-of-view (FOV). However, this makes it difficult to adjust the laser beam within the receiver's FOV. Since the same telescope is used to both transmit and receive the laser beam, a small amount of the emitted beam back-reflected from the beam splitter often damages the detector.

In this chapter, we describe a portable automated lidar (PAL) system, which we have developed to conduct full-time operation and all-weather observation through the laboratory window.[2–4] The PAL system has an automated correction mechanism for misalignment of the overlap between the transmitted laser beam and the receiver FOV. Hence the system is able to operate in a stable and stand-alone way. In addition, we have recently installed the scanning mechanism by attaching a horizontal stage to the PAL system. This improvement contributes greatly to monitoring the two-dimensional structure of the atmosphere nearly instantaneously.

2. PAL System

The PAL system is a variation of MPL system. The system configuration is shown in Fig. 1 and its specifications are summarized in Table 1. Since the transmitted energy is $15\,\mu$J, the system is nearly eye-safe at the expense of weak signals (lidar echo). To attain enough signal-to-noise ratio, the background light due to sky radiance must be eliminated with a narrow-bandwidth filter (0.5 nm) and a narrow FOV of 0.2 mrad. At the same time it is essential to keep the good overlap between the laser beam and the telescope FOV. Misalignment of the overlap, however, sometimes occurs from changes in the ambient temperature and accidental disturbances. The system has the auto-alignment mechanism, in which the laser beam is scanned vertically and then horizontally within the receiver's FOV and the maximum in the return signal (a certain range near the peak of the A-scope) is sought every 15 min.

The detector is a photo-multiplier operated in the photon counting mode (Hamamatsu photonics K.K. R1924P). The lidar echoes are

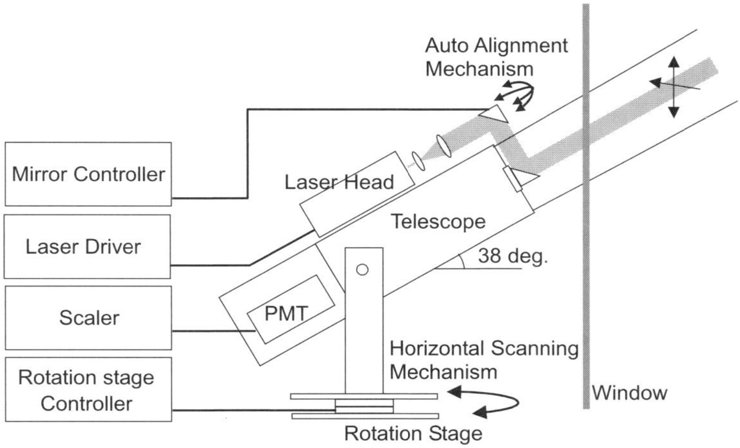

Fig. 1. System configuration of portable automated lidar.

Table 1. Specification of PAL.

Laser	LD pumped Nd:YAG Laser Pulse power 15 μJ Wavelength 532 nm
Detector	Photo-multiplier (photon counting mode)
Telescope	Schmidt-Cassegrain Aperture 20 cm diameter Field of View 0.2 mrad
Scaler	Resolution 24 m Range 24 m Averaging 10 or 20 s

accumulated by a scaler (Stanford Research Systems SR430). The spatial resolution is 24 m and the maximum observation range is 24 km (altitude 15 km). The observation is made though the vertical window of the laboratory, leading to the capability of measurement under all weather conditions. The observation data have been accumulated since the year 2004. The system status can be checked and the data can be downloaded through the Internet.

A built-in rotation stage for horizontal scanning has recently been installed. As the PAL system is fabricated as a monolithic structure including a laser head and a detector, the rotation stage was "inserted" under all the optical systems. The scanning observation of a range of ±25 degrees is conducted every hour, interrupting the continuous measurement for about 6 min. The PAL system is operated in Chiba Prefecture Environmental Research Center, with its beam pointed northward at the elevation angle of 38 degrees. The center is located on the east of Tokyo bay, about 10 km south of Chiba University. There is an industrial area and a busy load on the seaside (west of the center).

3. PAL Observation

The main advantage of the continuous and long-term observation is capturing the local weather change that takes place in a time scale of several hours. Especially, the system can monitor the onset and recovery of bad weather conditions and changes in polluted airs. These features are largely dependent on the site locations and conditions (urban/rural/mountains/waters). Two examples of characteristic results from the viewpoint of long-term cloud observation are shown in the following.

Figure 2 shows the result observed during 0–12 h local time on October 7, 2006. The weather map of Fig. 2(a)[5] shows that the low

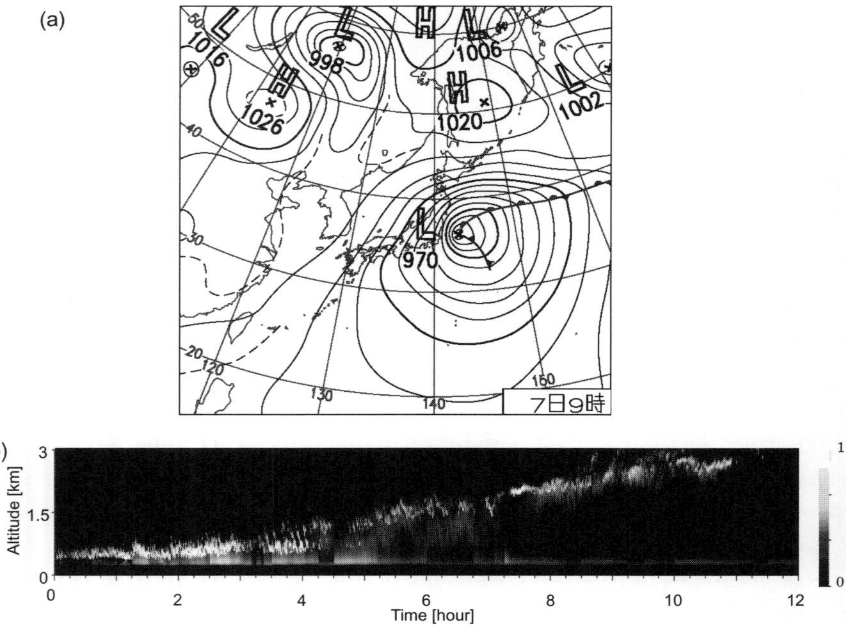

Fig. 2. (a) Weather map over Japan on October 7, 2006. (b) 12-h cloud long-term observation result: October 7, 2006. Temp. 21°C, Hum. 35%.

pressure has moved northward passing along the east coast of Japan, involving stationary and cold fronts. The PAL data in Fig. 2(b) also show that the long-lasting rain from the day before stopped and the cloud gradually gained altitude. The PAL data shown here are all corrected by the squared distance. Relatively, large echo appeared under the cloud till 7:00 (local time) in the morning. On that day, temperature and humidity largely changed at 7:00 (local time). Wind direction was northwest, and its speed was 8 m/s. The 10-h cloud elevation indicates the passage of highly developed low pressure. Figure 3 is the result observed during 0–12 on September 18, 2006. It was a windy day. Low clouds of less than 1 km altitude appeared during 0–5 h. They raised the altitude up to 2 km during 5–8 h. The lidar echo from these clouds was sparse and largely fluctuating in altitude. Sharp downturn of the cloud altitudes during 8–10 h was due to the rainfall.

Examples of long-term temporal motion of the atmosphere are shown in Fig. 4. Figure 4(a) is the result observed during 0–12 h local time on

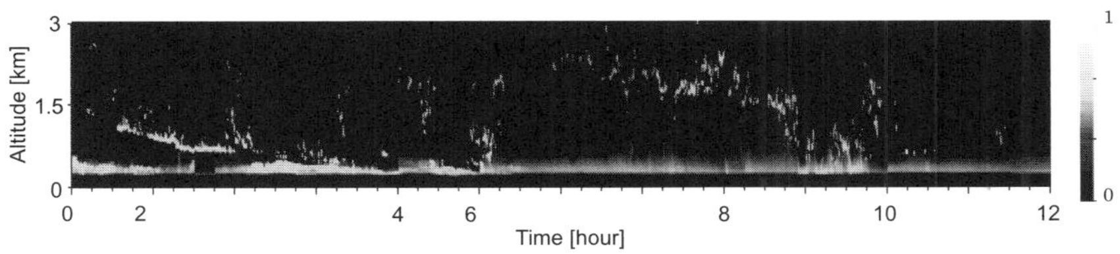

Fig. 3. 12-h cloud long-term observation result: September 18, 2006. Temp. 24°C, Hum. 82%.

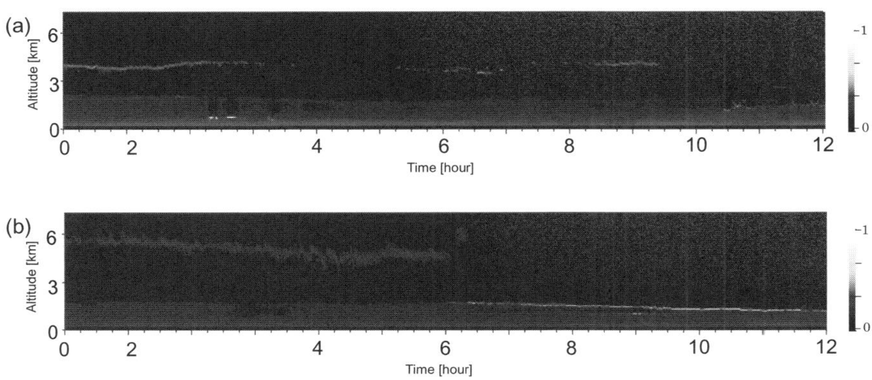

Fig. 4. 12-h atmosphere long-term observation results. (a) September 21, 2006. Temp. 24.7°C, Hum. 59%, (b) December 23, 2006. Temp. 11.9°C, Hum. 57%.

September 21, 2006. The atmospheric boundary layer and cloud were captured at the altitude of 2 km and 4 km, respectively. The structure was stable and showed little change till 9:00, while the relatively large echo appeared and raised its altitude from the ground during 9–12 h. In accordance with the change, cloud appeared at the altitude of 1.5–2 km. This condition continued till 16:00. Figure 4(b) shows the result observed during 0–12 h on December 23, 2006. The cloud appeared at the altitude of 6 km and lowered its altitude gradually from 0 to 6 h. Another cloud appeared on the boundary layer at the altitude of 1.5 km starting from 6:00. The boundary layer reduced the altitude down to 0.3–0.5 km. Furthermore during 8–10 h, another cloud appeared on the lowered boundary layer. Obviously, those results demonstrate the benefit of long-term observation. The change in temperature, wind, and the local-climatological influence of the site location will also be reflected in the observation data.

The result of temporal and horizontal-scanning observations on July 2, 2007 is shown in Fig. 5. On the day, the cloudy weather from the preceding day gradually worsened and started to rain in the evening. Time-height indication result of Fig. 5(a) shows that cloud moved slowly in the altitude range of 1–1.5 km during 0–8 h. The cloud altitude lowered in 8–12 h, while another thin echo appeared under the cloud layer. It rained in 15–19 h (Chiba city).

Temperature–humidity variation shown in Fig. 5(c) and pressure–wind speed variation in Fig. 5(d) also indicate the same change in the atmosphere activity, particularly the change of humidity in 0–12 h and 15–19 h, and the change of pressure/wind speed in 15–19 h. The spatial distributions of

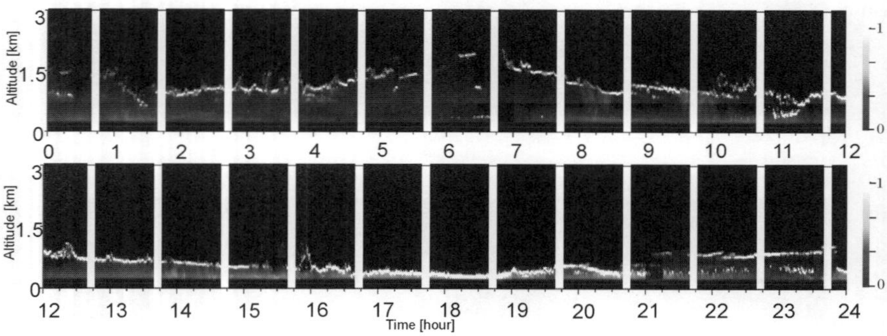

(a)

Fig. 5. (a) 24-h long-term observation result (b) 24-h horizontal scanning result (c) temperature and humidity (d) pressure and wind speed data: 2 July, 2007.

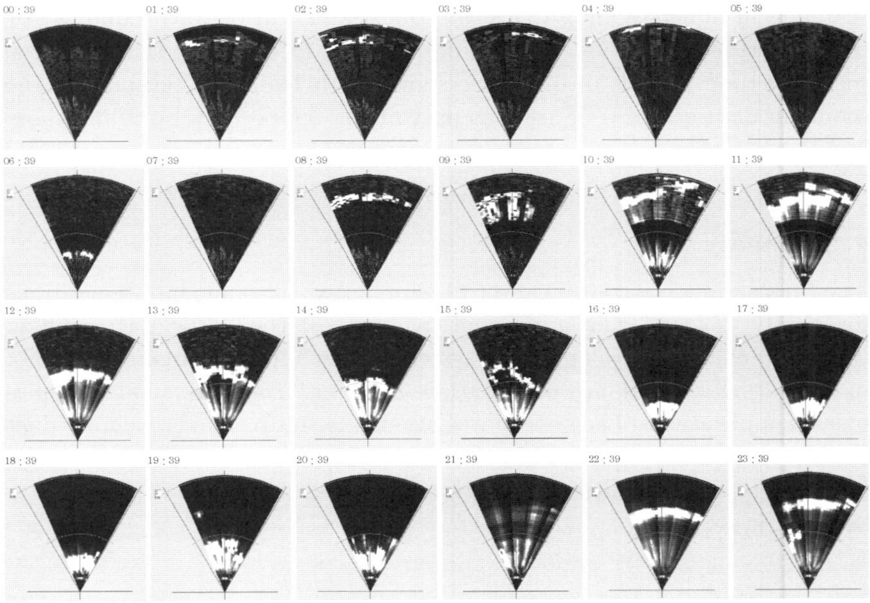

(b)

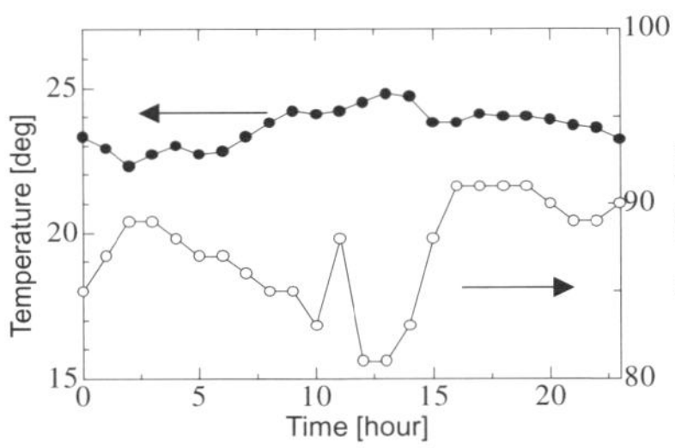

(c)

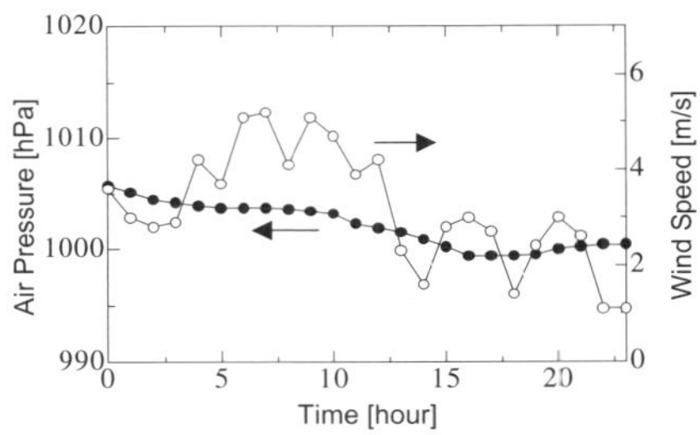

(d)

Fig. 5. (*Continued*).

the lidar echo obtained by the horizontal scanning are shown in Fig. 5(b). Although the scanning data are also corrected for the squared distance, it is not corrected for the elevation angle. Thus, the graphs are plotted in the beam propagation distance. The basic features of cloud echoes agree well with the temporal variation in Fig. 5(a), while the spatial structures of cloud are clearly detected in 9–12 h and 12–15 h by virtue of the horizontal scanning for the first time. The advantage of the horizontal scanning in understanding the local atmosphere will be fully exploited by deducing three-dimentional spatial information. In Fig. 5(a), the horizontal scanning time periods of 6 min are shown with blanks. Alternatively, the scanning data can also be used as part of the temporal data, filling those blanks.

4. Summary

The PAL system has continued the uninterrupted, autonomous observations for nearly 4 years. The additional inclusion of the horizontal scanning capability enables us to apply the system to new types of targets: spread of industrial smokes and dust distributions from busy roads are good examples of such applications. The system will also be useful to elucidate yellow dust activity and the pollen density distributions. In the near future, we are planning to install multi-wavelength and multi-polarization capabilities to the PAL system.

References

1. Spinhirne, Micro pulse lidar, *IEEE Trans. Geosci. Remote Sens.* **31**(1) (1993) 48–55.
2. N. Lagrosas *et al.*, Correlation study between suspended particulate matter and portable automated lidar data, *Aerosol Sci.* **36** (2005) 439–454.
3. G. Bagtasa, N. Takeuchi, S. Fukagawa, H. Kuze, T. Shiina, S. Naito, A. Sone and H. Kan, Mass extinction efficiency for tropospheric aerosols from potable automated lidar and β-ray SPM counter, *Proc. of 23rd International Laser Radar Conference* **3P-30** (2006) 499–502.
4. G. Bagtasa, C. Liu, N. Takeuchi, H. Kuze, S. Naito, A. Sone and H. Kan, Dual-site lidar observations and satellite data analysis for regional cloud characterization, *Opt. Rev.* **14** (2007) 39–47.
5. http://www.jma.go.jp/jma/indexe.html

Advances in Geosciences
Vol. 10: Atmospheric Science (2007)
Eds. J. H. Oh and G. P. Singh

SATELLITE-OBSERVED 3D MOISTURE STRUCTURE AND AIR–SEA INTERACTIONS DURING SUMMER MONSOON ONSET IN THE SOUTH CHINA SEA

YONGSHENG ZHANG
International Pacific Research Center, SOEST,
University of Hawaii at Manoa, POST Bldg. 401,
1680 East-West Road, Honolulu, Hawaii 96822, USA

TIM LI
International Pacific Research Center and Department of Meteorology,
SOEST, University of Hawaii at Manoa, POST Bldg 401,
1680 East-West Road, Honolulu, Hawaii 96822, USA

In this chapter, water vapor and air temperature profiles observed by the Atmospheric Infrared Sounder (AIRS), sea surface temperature (SST) and rain rate observed by TRMM Microwave Imager (TMI), and QuikSCAT surface wind for 2003–2006 are used to identify the 3D moisture structure and air–sea interaction processes during the onset of the South China Sea summer monsoon (SCSSM). Our analyses document an enhanced moisture accumulation in the atmospheric boundary layer co-existing with the surface easterlies preceding to the monsoon convection. Further analysis points out that, compared to the warming of SST, the boundary layer convergence plays a more important role in producing a warm and wet atmospheric boundary layer ahead of the monsoon convection, which contributes greatly to the development and maintenance of the northward propagation of the monsoon convection.

1. Introduction

As a semi-enclosed tropical sea surrounded by the Southeast–East Asian landmass, the South China Sea (SCS) plays an important role in modulation of climate anomalies in Asia. In middle May, accompanied by a switch of the prevailing zonal wind from easterly to westerly, the onset of the SCS summer monsoon (SCSSM) is characterized by an abrupt increase of precipitation and an associated tropical convergence zone northward propagating from the equator to northern SCS. The rainfall belt continues to move northward and controls the central China and South of Japan in late May and early

June.[1–3] On an interannual time-scale, the year-to-year variability of the SCSSM onset date and intensity in May has a strong projection onto the summer rainfall anomalous pattern in the East China and West Pacific Ocean, foreshadowing the development of the full-scale Asian summer monsoon during the subsequent months.[2,4–8] These scientific issues have called for a multi-national atmospheric and oceanographic observational and research plan, the SCS Monsoon Experiment (SCSMEX), which was aimed to a better understanding of the onset, maintenance, and variability of the SCSSM (for a overview, see Ref. 7).

In the past decades, efforts have been made to explore various aspects of the SCSSM and led to significant progress. However, some open issues still remain. For example, what is the driving mechanism of the SCSSM onset? While many studies focused on the large-scale environmental condition favorable for the SCSSM onset, the role of the local air–sea interaction and the three-dimensional water vapor profile has rarely been addressed, partially due to shortage of reliable observations in an appropriate time–space resolution. Using ship observations, Chu and Chang[9] identified the development of a warm-core eddy in the central SCS immediately before onset of the SCSSM attributing to the radiative warming and the downwelling driven by the surface anti-cyclonic flows, which helps lowering atmospheric surface pressure. However, their data analysis is limited to 1966 only. The air–sea heat exchanges during different stages of the SCSSM onset have also been explored by using the station observations in the SCS.[10–12] Though considerable air–sea flux exchanges were identified during the monsoon onset, the direction of the heat transportation varies from one study to another, partially because of the difference of the observation location and time.[12] So far there is no conclusive result on how the air–sea interaction and water vapor profile may affect the *in situ* thermodynamic condition which leads to the onset and northward propagation of the SCSSM.

The recently available satellite observations of the sea surface temperature (SST), precipitation, humidity, air temperature, and surface wind provide accurate and high-resolution coverage in the ocean regions where the conventional observation is rare. This provides an unprecedented opportunity to investigate the complicated physical processes relevant to the SCSSM onset. Among satellite sensors, the Atmospheric Infrared Sounder (AIRS) is a facility instrument aboard the NASA's Earth Observing System (EOS) polar-orbiting platform and is the most advanced moisture and air temperature sounding system. It constitutes an innovative atmospheric sounding group of visible, infrared, and microwave sensors for measuring atmospheric water and temperature profiles with a twice-daily, 1–2 km vertical, and 45 km horizontal resolutions. The accuracy of the humidity and air temperature profiles derived from AIRS have been recognized as improving forecasts from meteorological prediction models.[a] The advantages of using the AIRS in describing the air temperature and moisture structures of the tropical Madden–Julian oscillation (MJO) have been demonstrated by a couple of studies.[13] (Yang *et al.*, 2006). The instruments carried by the Tropical Rainfall Measuring Mission (TRMM) satellites provide useful information of the tropical rain rate and SST. Also, NASA's Quick Scatterometer (QuikSCAT) offers the information of the surface wind.

In general, the goal of this chapter is to reveal the role of *in situ* hydrological cycle in driving the northward movement of the tropical convection during the SCSSM onset by analyzing the three-dimensional water vapor and the underlying air–sea interaction using the aforementioned satellite observations during 2003–2006.

2. Datasets

The level-3 AIRS data used in this study include the atmospheric moisture and temperature profiles at 12 levels from 1000 to 100 hPa with a spatially 1.0 degree longitude–latitude and a temporally twice-daily resolutions since 1 August 2002. Detail description of this dataset can be obtained at: http://disc.sci.gsfc.nasa.gov/AIRS. In this chapter, 10-day mean data are constructed from the original twice-daily data. We also used the 3-day running mean rain rate and SST observed by the TRMM Microwave Imager (TMI) and surface wind observed by the QuikSCAT. Both TMI rain rate, SST, and QuikSCAT wind have a resolution of 0.25×0.25 longitude/latitude and the information is provided at: http://www.ssmi.com. Other data used in this study include the daily reanalysis of the National Centers for Environmental Prediction (NCEP) and the National Center for Atmospheric Research (NCAR), and the global surface and upper air analyses at the European Centre for Medium-Range Weather Forecasts (ECMWF). The later is a 2.5×2.5 degree grid output from the ECMWF operational model provided through NCAR.

---

[a]NOAA administrator Lautenbacher has reported that "the AIRS instrument has provided the most significant increase in forecast improvement in this time range of any other single instrument." (http://daac.gsfc.nasa.gov/AIRS/).

3. Results

Previous studies have documented the weakness of the humidity in the reanalyses, particularly, at the lower troposphere. Zhang[14] compared the monthly humidity from the NCEP/NCAR and ECMWF 40-year reanalyses with the station observations in the East China in 1990s, and he found that the humidity in both the NCEP/NCAR and ECMWF 40-year reanalyses in the lower troposphere is much larger than that from the station observation, concurrent with a cold bias in the air temperature. Tian *et al.*[15] identified that the lower-troposphere moisture and temperature structure related to MJO is much less well defined in NCEP than in AIRS in the Pacific and Indian Oceans. In Fig. 1, we compare the 10-day mean humidity at 1000 hPa in May for 2003–2006 from AIRS observations with the NCEP/NCAR reanalysis. Overall, the magnitude in NCEP/NCAR reanalysis is larger than those of AIRS observations. In the SCS region, the moisture maximum in the AIRS observation locates in the ocean but in the NCEP/NCAR

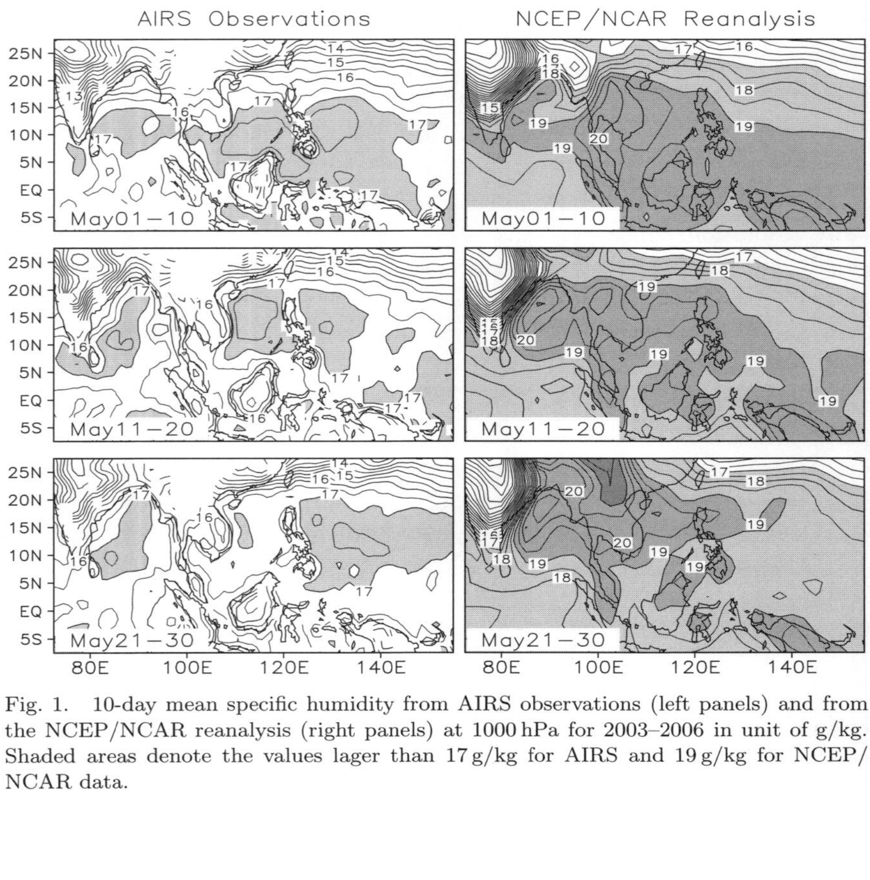

Fig. 1. 10-day mean specific humidity from AIRS observations (left panels) and from the NCEP/NCAR reanalysis (right panels) at 1000 hPa for 2003–2006 in unit of g/kg. Shaded areas denote the values lager than 17 g/kg for AIRS and 19 g/kg for NCEP/NCAR data.

reanalysis it mainly appears in the land region. In the northwestern Pacific region to east of Philippines, the AIRS observations show an independent maximum, but this does not occur in the NCEP/NCAR reanalysis. The results from AIRS observations also show a remarkable sub-seasonal change with decrease of the humidity from early to late May, consistent with a low-level moisture loss associated with a development of the strong convection activities in the SCS. However, this feature does not appear in the NCEP/NCAR reanalysis.

The 3D structure and evolution characteristics of the water vapor during the SCSSM onset have not been well documented. Figure 2 depicts the vertical distributions of the humidity, the QuiSCAT surface wind, and TMI precipitation averaged between 105°E–120°E in early, middle, and late May in 2003, 2004, and 2005, respectively. We did not discuss the case of 2006 simply because the onset of SCSSM in 2006 is strongly controlled by the circulation associated with Typhoon CHANCHU in middle May.

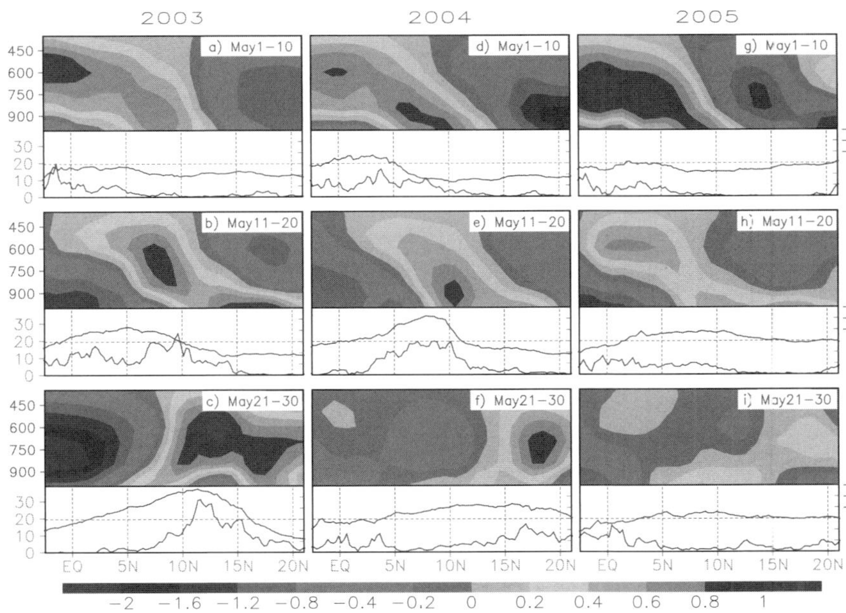

Fig. 2. The vertical-latitude distributions of the 10-day mean AIRS humidity at 1000 hPa (shading, scales are shown in the bar in unit of g/kg), QuiSCAT surface zonal wind (blue line, Y-coordinate scales are marked in right side in unit of m/s) and TMI rain rate (red line, Y-coordinate scales are marked in left side in unit of mm/day) averaged between 105°E and 120°E. For the humidity, Y-coordinate shows the pressure level (hPa) and the domain average over 105°E–120°E and eq.-20°N is removed at each pressure level.

Generally, the onset of the SCSSM is accompanied by a well-defined vertical southward-tilting moisture structure in the southern and central SCS in early and middle May (Fig. 2). In 2003, the enhanced convective activities occur in middle May in the central SCS around 12°N–15°N (Fig. 2(b)), signaling the onset of the SCSSM. Then, the rainfall maximum moves northward in late May (Fig. 2(c)). In 2004, strong rainfall starts in the southern SCS in early May (Fig. 2(d)), and then the rainfall maximum moves northward in middle and later May (Figs. 2(f) and (g)). In 2005, the convective activities in May are much weaker compared to those in 2003 and 2004, but the northward propagation of the convective activities from the southern to northern SCS is still seen clearly.

The most important feature depicted in Fig. 2 is that the rainfall maximum is preceded by water vapor maximum in the atmospheric boundary layer (ABL, around 900–1000 hPa) and is followed by a dry phase. The former is overlapping with easterlies, and latter with westerlies. In 2003, the vapor maximum in the low-to-middle troposphere (850–500 hPa) is clearly concurrent with the rainfall maximum (Figs. 2(a) and (b)). But near the surface the water vapor maximum occurs in the front of the rainfall maximum and is confined in the easterly prevailing region. The feature can also be clearly identified in 2004 and 2005 (Figs. 2(d), (e), (g), and (h)). This indicates that the enhanced moisture accumulation in the ABL in front of the monsoon convection plays an active role in leading the northward movement of the convection.

In order to identify what causes an enhanced moisture accumulation in ABL in front of the monsoon convection, we plot the south–north distributions of the AIRS humidity at 1000 hPa, TMI SST, ECMWF analyzed latent heat flux, QuikSCAT wind divergence, and TMI rain rate along 105°E–120°E during early, middle, and late May. Figure 3 presents the result in 2003 and the following features are noteworthy:

(1) The 1000 hPa humidity experiences a sharply decrease from early to middle May (Fig. 3(a), thin solid and dashed lines), when strong monsoon rainfall occurs in the southern SCS (south of 10°N). But it recovers in late May (Fig. 3(a) thick solid line), as the monsoon rainfall moves into the northern SCS with maximum in the region of 10°N–15°N. In the northern SCS (north to 10°N), the decrease of humidity in ABL is coincident with a maximum rainfall moving from the southern SCS (Figs. 3(a) and (e), thick solid line).

(2) A decrease of SST occurs from early to late May in the southern SCS is concurrent with a rapid increase of SST in the northern SCS

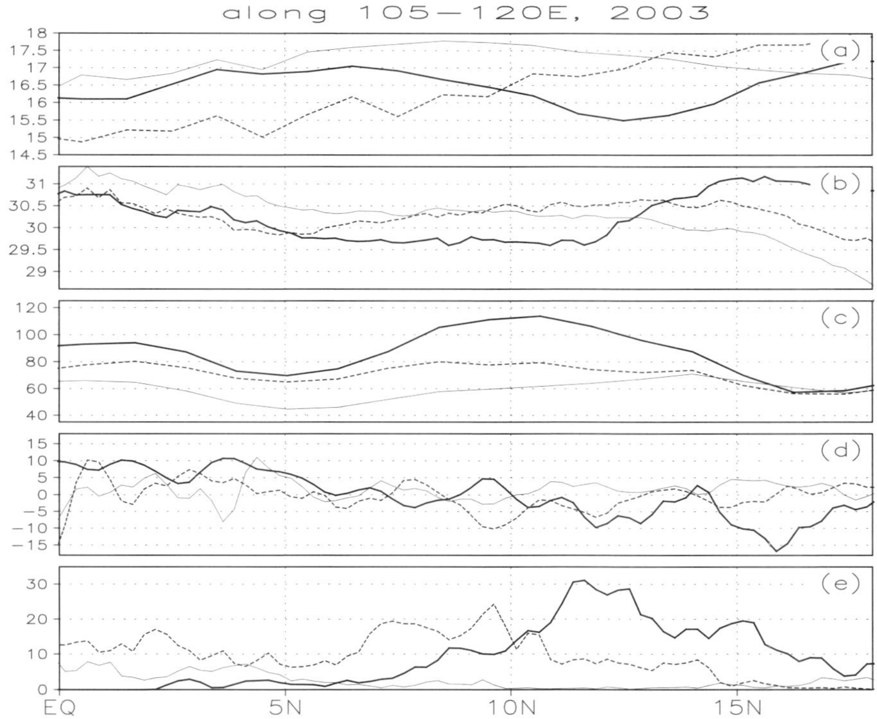

Fig. 3. The 10-day average of south–north distributions of (a) the AIRS humidity at 1000 hPa (g/kg), (b) SST (°C), (c) latent heat flux from the ocean to the atmosphere (w/m^2), (d) divergence calculated from the QuikSCAT surface wind (10^{-6} s), and (e) TMI rain rate (mm/day) along 105°E–120°E during May 1–10 (solid thin line), May 11–20 (dashed line), and May 21–30 (thick solid line) in 2003.

(Fig. 3(b)). This implies that the ocean and atmospheric physical processes associated with development of the strong monsoon rainfall tend to decrease the underlying SST, but to warm the SST in front of the rainfall maximum.

(3) The decrease of the moisture in ABL in the southern SCS in middle May (Fig. 3(a), dashed line) and in the northern SCS in late May (Fig. 3(c), solid thick line) are concurrent with an increase of the latent heat flux (Fig. 3(c)) relevant to a burst of westerly in these two regions (Figs. 2(b) and (c), blue lines), indicating that the moisture change at ABL is not attributed to the surface evaporation.

(4) While a strong convection maximum co-locates with a strong convergence near the surface, convergence also occurs in front of the

rainfall maximum, tending to overlap the region of enhanced moisture (Fig. 3(d)).

Figure 4 shows the same plot for 2004. The strong monsoon rainfall first appears in the southern SCS in early May and then moves to northern SCS. In middle May, the development of strong monsoon rainfall in the central SCS (Fig. 4(e), dashed line) is concurrent with a reduced humidity (Fig. 4(a), dashed line) and SST (Fig. 4(b)), but an increased latent heat flux transporting from the ocean to the atmosphere in the southern SCS (Fig. 4(c)). The convergence mainly locates to the front of the rainfall maximum (Fig. 4(d)). Similar to 2003, a decrease of SST in the region south to 10°N is accompanying with an increase in the northern part. In late May, a pronounced decrease of SST in the central–northern SCS coincides with

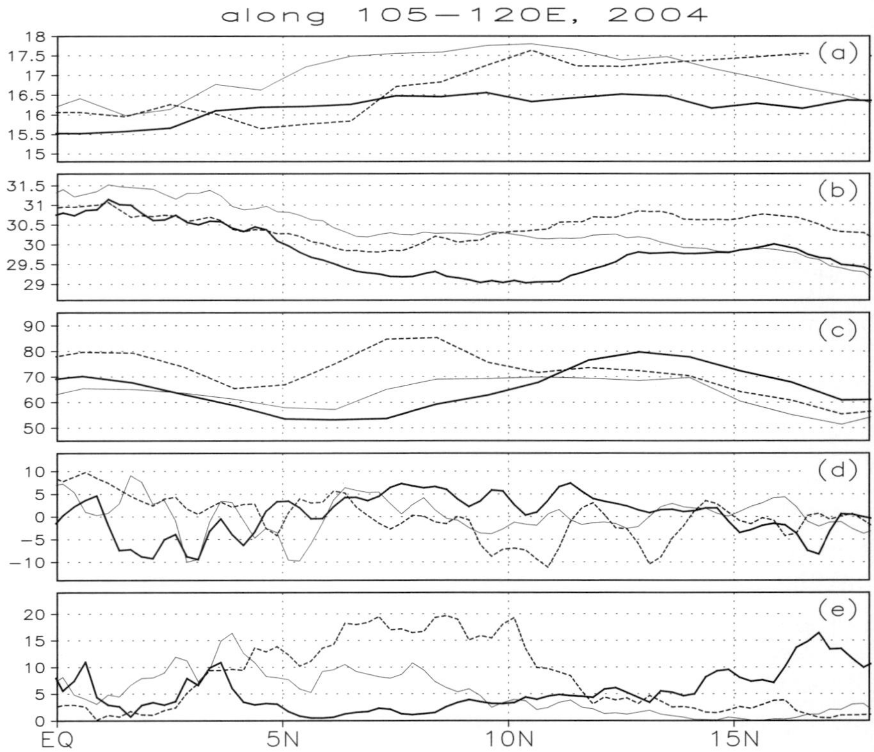

Fig. 4. Same as Fig. 3, but for the year of 2004.

an increase of the latent heat flux in the northern SCS, just behind the rainfall maximum.

It is interesting to note that, in Figs. 3 and 4, the maximum of evaporation reflected by the latent heating flux is not coincident with the enhanced moisture accumulation in front of the intense convection in early and middle May. In both 2003 and 2004, while a remarkable increase of SST and moisture occurs in the northern SCS in middle May (Figs. 5(a) and (b)), the latent heat flux has a minimum change (Fig. 5(c)). The maximum increase in latent heat flux between 5°N and 10°N is primarily due to the increase in the surface wind speed (Fig. 5(d)), and it coexists with a SST cooling (Fig. 5(b)) and a sharp decrease of the humidity. When the latent heating flux maximum moves into the northern SCS in late May in 2003

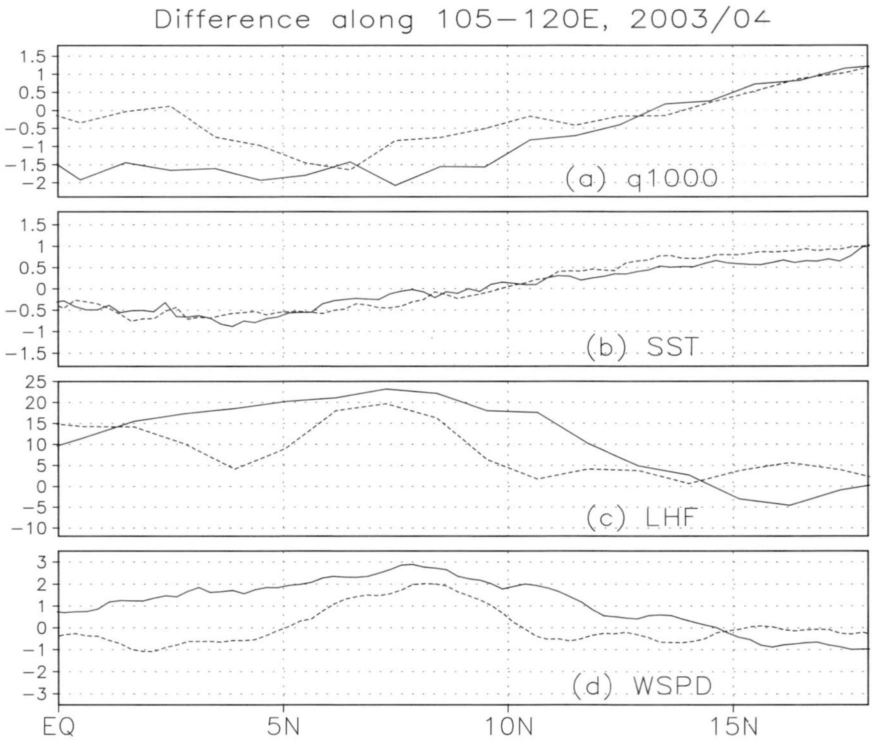

Fig. 5. The difference of 10-day average between May 11–20 and May 1–10 in 2003 (solid lines) and 2004 (dashed lines) of (a) the AIRS humidity at 1000 hPa (g/kg), (b) SST (°C), (c) latent heat flux from the ocean to the atmosphere (w/m^2), and (d) wind speed (m/s) along 105°E–120°E.

and 2004, the humidity near the surface in the northern SCS decreases. The results indicate that, while a SST warming is concurrent with a moistening ABL in front of the monsoon convection, this warming does not directly contribute to the moistening. Above features can also be found in the year of 2005 though, compared to 2003 and 2004, the strength of the monsoon rainfall and associated wind are not as strong as those in 2003 and 2004.

As shown in Figs. 3(d) and 4(d), lower-level convergence not only coincides with the convection but also appears in front of the rainfall maximum and is co-located with the region of maximum boundary-layer humidity. This suggests that the convergence in the ABL is one of dominant

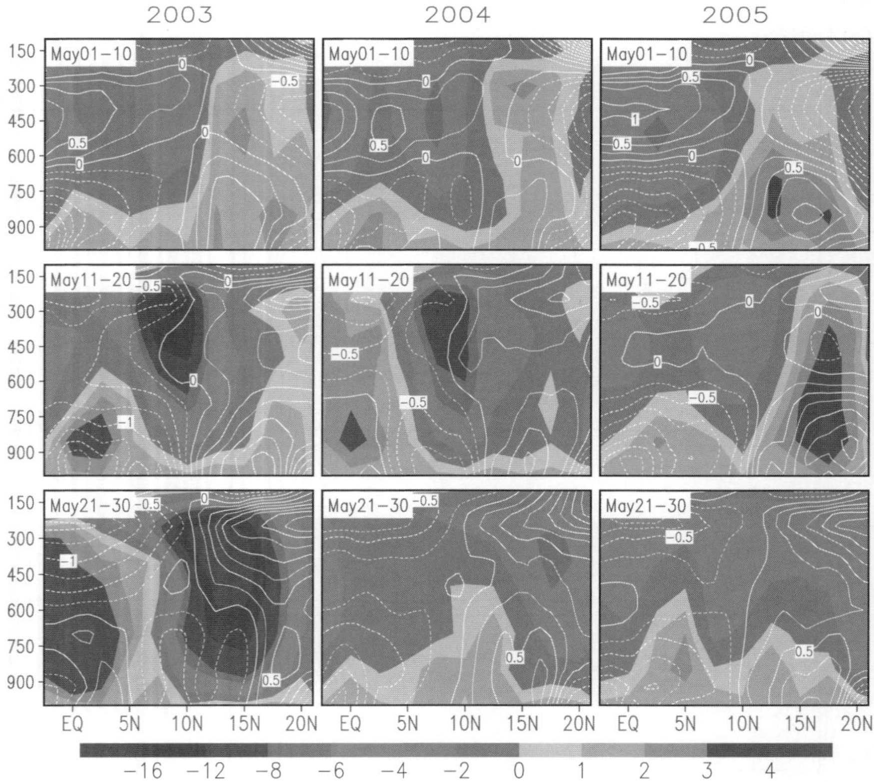

Fig. 6. The latitude-pressure cross-sections of the ECMWF vertical velocity (Omega, shading and scales are shown by the bar in unit of 10^{-2} Pa/s) and AIRS air temperature (contours, with an interval of 0.25°C) averaged along 105°E–120°E. A domain average over 105°E–102°E and eq-20°N is removed from the air temperature at each pressure level.

processes responsible for an enhanced vapor accumulation in front of the convection.

Figure 6 presents the latitude-pressure cross-sections of the vertical velocity (Omega) from ECMWF analysis and the air temperature observed by AIRS averaged along 105°E–120°E. A domain average over 105°E–102°E and eq.-20°N is removed from the air temperature at each level in order to emphasize its north–south gradient. It can be seen that, during monsoon onset in 2003, 2004, and 2005, the northward movement of the advanced convection in Fig. 2 (red lines) is associated with strongly ascending motion between 600 and 200 hPa (Fig. 6, shading). In the pre-onset phase during early May, a weak descending branch is found in front of the convection maximum. During the mature onset phase (middle May), the descending motion weakens and is confined below 700 hPa in 2003 and 2004. The weak descending motion in front of the monsoon convection helps to preserve the converged moisture near the surface. The air temperature profile observed from AIRS satellite show that a relatively warm center lies in front of the rainfall maximum while a cold one overlapping the rainfall maximum. Therefore, the monsoon rainfall maximum is overlapped and followed by a dry-cold ABL and preceded by a warm–wet one. This provides a favorable condition for the northward migration of the convection from southern to northern SCS.

4. Summary and Discussion

Using the high-resolution moisture and air temperature data observed by the AIRS, tropical rain rate and SST by TRIMM and QuiSCAT surface wind field for 2003–2005, we investigate the 3D moisture structure and air temperature, and air–sea interactions involved in northward movement of the strong convective activities during the summer monsoon onset in the SCS. Our special attention is paid to reveal how the regional circulation related to the monsoon onset in the SCS contributes to the *in situ* air–sea interactions and moisture distributions which, in turn, provide a favorable condition for development and maintenance of the tropical convective activities, and the northward propagation of the monsoon rainfall. The results show that the SCSSM onset in 2003, 2004, and 2005 shows some common features. Most interestingly, we find that the intense convective activities are preceded by an enhanced moisture accumulation and warm air temperature near the surface and followed by a dry and cold ABL. The vertical profile shows a southward-tilting structure. The wet ABL in front of the monsoon rainfall maximum is co-located with easterly wind, and the maximum of the latent heat flux transporting from the ocean to the atmosphere concurs with the burst of the westerly behind the convective maximum and tends to coexist with a SST cooling and a water vapor decrease near the surface. Therefore, despite the in-phase relation between the enhanced moisture accumulation and increased SST in front of the monsoon rainfall maximum, the SST warming and associated evaporation do not directly contribute to the humidity increase.

Our diagnosis also shows that the warm and wet ABL in front of the monsoon rainfall maximum tends to be concurrent with a near-surface convergence. Therefore, it is likely that the lower-level convergence is the major contributor for the moisture increase. In contrast to Fu *et al.*,[13] our study emphasizes the role of the regional convergence preceding the monsoon rainfall in determining the enhanced moisture accumulation. This result is in agreement with the observational and theoretical study by Jiang *et al.*,[16] who pointed out that the preceding distribution of the boundary-layer convergence is one of the mechanisms that contributes to a northward propagation of the intraseasonal convection.

A special feature of the current study is to reveal a more conclusive and systematical scenario on the regional feedback between the atmospheric moisture, air temperature, and SST response to the intense monsoonal convections with the aid of the fine-resolution satellite observations from AIRS, TMI, and QuikSCAT instruments. More detail linkage needs to be further identified with the help of the ocean data and an accurate estimation of the surface heat fluxes. In these feedbacks, we observe that the warm and wet ABL provides more unstable condition preceding the convective maximum, and strong temperature gradients associated with warming (cooling) in front of (behind) the convection, in turn, make a great contribution to the northward propagation of the monsoon convective activities.

Acknowledgments

TL was supported by ONR grants N000140710145 and N000140210532 and NRL subcontract N00173-06-1-G031. The International Pacific Research Center is partially sponsored by the Japan Agency for Marine-Earth Science and Technology (JAMSTEC).

References

1. S. Y. Tao and L. X. Chen, *Monsoon Meteorology* (Oxford University Press, 1987), p. 60.
2. K. M. Lau and S. Yang, *Adv. Atmos. Sci.* **14** (1997) 141.
3. B. Wang and L. Ho, *J. Clim.* **15** (2002) 386.
4. C. H. So and J. C. L. Chan, *J. Meteor. Soc. Jpn.* **75** (1997) 43.
5. B. Wang and R. Wu, *Adv. Atmos. Sci.* **14** (1997) 177.
6. C. Li and L. Zhang, *Chinese J. Atmos. Sci.* **23** (1999) 257.
7. Y. Ding, C. Li and Y. Liu, *Adv. Atmos. Sci.* **21** (2004) 343.
8. R. Huang, L. Gu, L. Zhou and S. Wu, *Adv. Atmos. Sci.* **23** (2006) 909.
9. P. C. Chu and C.-P. Chang, *Adv. Atmos. Sci.* **14** (1997) 195.
10. X. Bai, A. Wu and Y. Zhao, *Onset and Evolution of the South China Sea Monsoon and Its Interaction with the Ocean* (China Meteor. Press, 1999), p. 381.
11. J. Yan *et al.*, *Acta Ocea. Sin.* **22** (2004) 369.
12. D. Wu *et al.*, *Chinese Sci. Bull.* **51** (2006) 2413.
13. X. Fu, B. Wang and L. Tao, *Geophys. Res. Lett.* **33** (2006) L03705.
14. Y. Zhang, *Biases of NCEP/NCAR and ECMWF 40-Year Re-Analyses in East Asia*, IAMAS 2005 Scientific Assembly, August 2–11, 2005, Beijing, China.
15. B. Tian *et al.*, *J. Atmos. Sci.* **63** (2006) 2462.
16. X. Jiang, T. Li and B. Wang, *J. Clim.* **17** (2004) 1022.

Advances in Geosciences
Vol. 10: Atmospheric Science (2007)
Eds. J. H. Oh and G. P. Singh

EAST ASIAN SUMMER MONSOON AND THE RAINFALL IN EAST CHINA

LU XINYAN*,†,‡, ZHANG XIUZHI§ and CHEN JINNIAN*

*Institute of Oceanology, Chinese Academy of Sciences, Qingdao 266071, China
‡lvxy@cma.gov.cv

†Graduate School of the Chinese Academy of Sciences, Beijing 100049, China

§Beijing Climate Center, Beijing 100081, China

Using daily rainfall measurements from 740 stations across China and European Center for Medium-Range Weather Forecasts (ECMWF) upper air reanalysis daily data (1958–2001), we give out climatically characters of East Asian Summer Monsoon's (EASM) movement with the definition of the EASM's front, finding out that the transfer of the rain belt over East China is consistent with the advance and retreat of the EASM. By the empirical orthogonal function (EOF) analysis of the gridded EASM's index (average for the 28th–45th pentad) from 1958 to 2001 in the area (105°E–150°E, 15°N–55°N), it is founded that, the second mode of the EOF analysis exhibits interdecadal variations and indicate that the movement of EASM has three interdecadal abrupt changes in 1965, 1980, and 1994, respectively. Therefore, the three interdecadal abrupt changes bring the different processes of the EASM's movement and lead to the obvious change of the spatial distribution pattern of summer rainfall in East China directly, especially prior to 1965, the rainfall in the mid-lower reaches of the Yangtze River is much less than normal, while the precipitation is much more in South China, North China, and Northeast China but decreasing continuously since 1965. However, the rainfall in the mid-lower Yangtze Valley increases continually from 1980, especially from 1994 the rainfall in South China and the Yangtze Valley increases rapidly while the precipitation over North China was much less than normal. Therefore, East China underwent from the pattern of south-drought and northern-waterlog before 1979 to south-waterlog and north-drought.

1. Introduction

China is in a typical monsoon zone and the East Asian Summer Monsoon (EASM) controls the distribution of most rainfall over China from June to

‡Corresponding author.

August every year, the transfer of rain belt and the most of drought and floods in China, especially in the East China. After the onset of EASM, Pre-flood Season Precipitation reaches to be strongest period, and then the major seasonal rain belt over China extend northward to the Yangtze–Huaihe River Valley, North China, and Northeast China, respectively.[1,2] The stay duration and the strength of EASM in its journey to the north determine directly local rainfall amount on its way, however, as the intensity of monsoon and the process of advance and retreat of EASM are different every year, so the precipitation of different regions in East China has the visible inter-annual variability, and this is also the important natural cause of the high-frequency and fierce disaster, such as the drought, flood, clod, and hot. Therefore, describing and monitoring the movement of EASM is important value to the prediction of the rain belt.

Huang and Zhou[3] showed that the interdecadal variability of EASM is very obvious during the recent 50 years. EASM is stronger than normal from 1950 to mid-1960s, while it turns to be weaker since mid-1960s. Based on Shi Neng's summer monsoon index (SMI), Zhang *et al.*[4] also indicated that EASM is continually stronger from 1950 to mid-1960s and it became weaker from mid-1960s, especially since 1980 EASM became weaker and weaker. Guo *et al.*[5] pointed out that there is a systematic reduction of EASM during the period of 1951–2000 and EASM have obvious interdecadal variability. Strong monsoon (SMI = 1.0) was predominated during the period of 1951–1975; SMI was less than 1.0 since 1976. Wu *et al.*[6] calculated the location of northward shift of EASM by using NCEP/NCAR air temperature and specific humidity reanalysis data in recent 50 years. They found that there is obvious interdecadal change in the north boundary of EASM, and the years with the far north boundary centralize in 1950s and 1960s, with the far south in 1980s and 1990s. Wu and Qian[7] defined the north edge of EASM based on precipitation, wind and pseudo-equivalent potential temperature, basing the northernmost location that the EASM can reach from 1961 to 2001. They pointed out that the northernmost location that the EASM can reach have dominant interdecadal shift during the period of 1977–1979. Kwon *et al.*[8] showed that the general circulation over East Asia has a shift during mid-1990s.

Wang *et al.*[9] showed that the rainfall in East China exhibits obvious interdecadal variability. The temporal and spatial distribution of the summer precipitation in East China has some regularity. Zhao[10] divided the distribution of summer precipitation in East China into three different patterns: the northern type, the middle type, and the southern type.

Huang *et al.*[11,12] pointed that the summer rainfall in China exhibits two abrupt changes of climate which occurred in 1965 and 1976 respectively, the summer precipitation over North China begins to decrease obviously from 1965, but the rainfall in the middle-lower Yangtze River — the Huai River Valley keeps increasing since the later 1970s, while the precipitations in South and North China in the period from 1980s to the early 1990s were obviously less than those in the 1970s and drought trend was more and more severe there. However, from the mid-1990s, there was an increasing trend in the precipitation in the northern part of North China. Xu and Wei[13] showed that there are four periods in the precipitation of North China. The plenitude of precipitation appears from 1880 to 1898 and 1949 to 1964 and the low rain water appears from 1899 to 1947 and from 1965 to 1999. Shi and Xu[14] showed that the precipitation over China exhibits the transition of interdecadal trend occurs in 1980s by using mode of the trend discretion, the rainfall in south of China increases after the transition, but rainfall in north of china begins to decrease.

From the above analysis, we know that the EASM exhibits distinct interdecadal variability. The before study mainly bases on two methods. The first method of EASM index is used by Zhang Zhixiu, Guo Qiyun and so on. They showed that EASM experiences interdecadal variability by the analyses of time series of EASM's index. But these indices were developed for EASM as a whole on a monthly or seasonal basis. In addition, Wu Changgang, Hu Haoran, Qian Weihong *et al.* used the variations of EASM's north edge to discuss the interdecadal variability. In fact, EASM moves northward from low to middle and high latitudes or retreat southward as the EASM develop and it do not stay at only one region with less movement. Therefore, it brings different influence to different regions. A single index or the north edges fail to describe the condition during EASM's advance/retreat and its vigor locally. The study about how the advance and retreat of the EASM influence the drought and flood is much less, especially in interdecadal timescales. In addition, we know that the precipitation over East China also has interdecadal variability and the spatial distribution pattern of the summer fall in East China has distinct transition. Whether the EASM's movement also exhibits interdecadal variability? How is the relationship between the interdecadal variability of EASM's movement and the spatial distribution pattern of the rainfall in East China?

In order to answer above questions, by using ECMWF $2.5^\circ \times 2.5^\circ$ upper air reanalysis daily data and 740 stations rainfall data compiled by the National Meteorological Information Center of China (1958–2001), and based on the EASM's gridded index and the definition of EASM's front,[15] we analyzed the climatological characters of EASM's movement and the relationship between the EASM's movement and the spatial distribution pattern of the summer rainfall over eastern China in interdecadal time-scales.

2. The EASM's Movement and Rain Belt in Eastern China

Based on the previous studies, Lu *et al.*[15] proposed the definition of EASM's gridded index by using low-level SW wind and specific humidity, i.e.

$$\mathrm{Im} = \frac{V_{\mathrm{sw}} - \overline{V_{\mathrm{sw}}}}{\delta_{\mathrm{sw}}} + \frac{Q - \overline{Q}}{\delta q}, \tag{1}$$

where V_{sw} is the projection of total wind speed on the SW direction at a 850 hPa for a grid point, Q is the mean specific humidity at 925 and 850 hPa for a grid point; $\overline{V_{\mathrm{sw}}}$ is the projection of 850-hPa total wind speed on the SW direction, averaged over (10°N–50°N, 90°E–150°E); $\overline{Q}$ is the mean specific humidity at 925 and 850 hPa over the same area; δ_{sw} and δq are the standard deviation of V_{sw} and Q in the above region. In the process of the EASM's movement, the $\mathrm{Im} = 1.0$ ($= 0.3$) line south (north) of 35°N is denoted as its leading edge. It is proved that the defined EASM's leading edge (or front) can describe the movement of the EASM objectively.

Using the above definition, we get the 44-year average process of the EASM's movement (1958–2001) (Fig. 1). Based on the advance and retreat

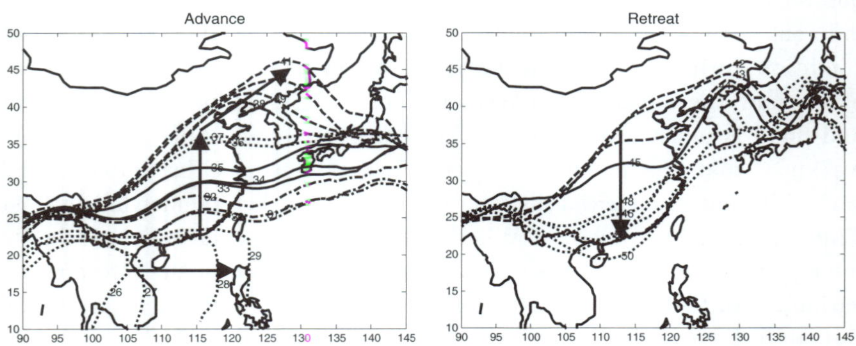

(a) (b)

Fig. 1. The movement of EASM. (a) The advance, (b) the retreat (the number in the pictures refer to the time (pentad)).

of EASM, we divided the process of EASM's movement into the following five phases:

1. *The EASM's onset and its front concentrates in South China*

On average, the Asian summer monsoon breaks out in SCS at 28th pentad. During the 28th and 29th pentad, the front stays at South China Sea (SCS) and nearby Philippine and moves northward and eastward continually. The rain belt over East China is influenced by the southwesterly flows from SCS and the cold air from mid-high latitude and start retreat southward to South China, but the rainfall over South China increase rapidly. As the strong westerly flows from the Indian Ocean and southeasterly flows from the south of West Pacific subtropical high meet in SCS and then they move northward during the 30th–32nd pentad, the front of EASM reaches South China and stays there steadily (Fig. 2(a)). Meantime, Pre-flood Rainy

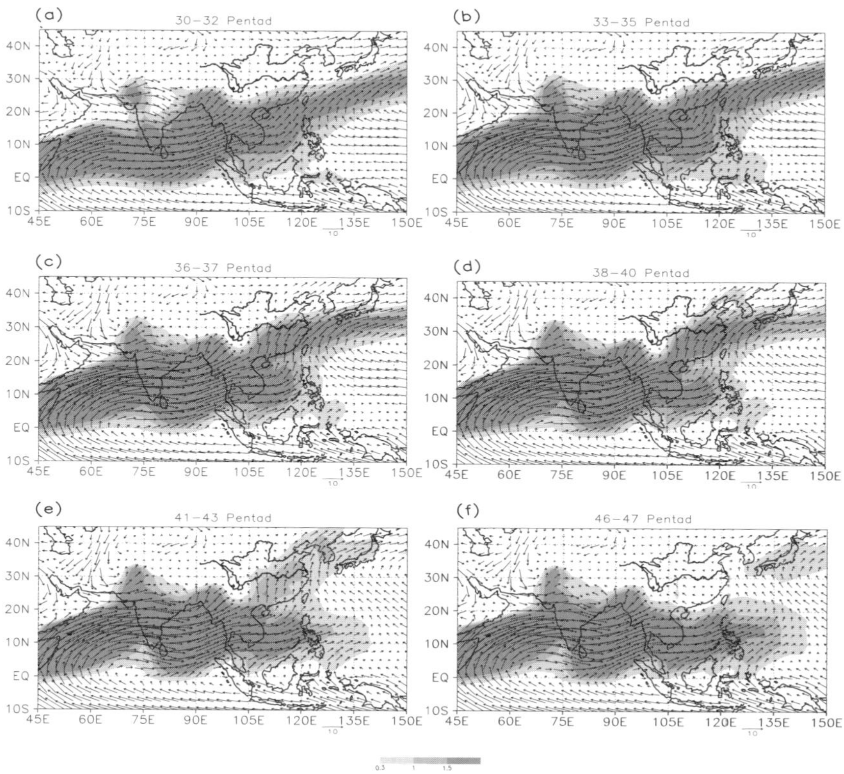

Fig. 2. The distribution of the mean wind (850 hPa) and EASM's index averaged for 1958–2001. (a) 30th–32nd Pentad, (b) 33rd–35th Pentad, (c) 36th–37th Pentad, (d) 38th–40th Pentad, (e) 41st–43rd Pentad, and (f) 46th–47th Pentad.

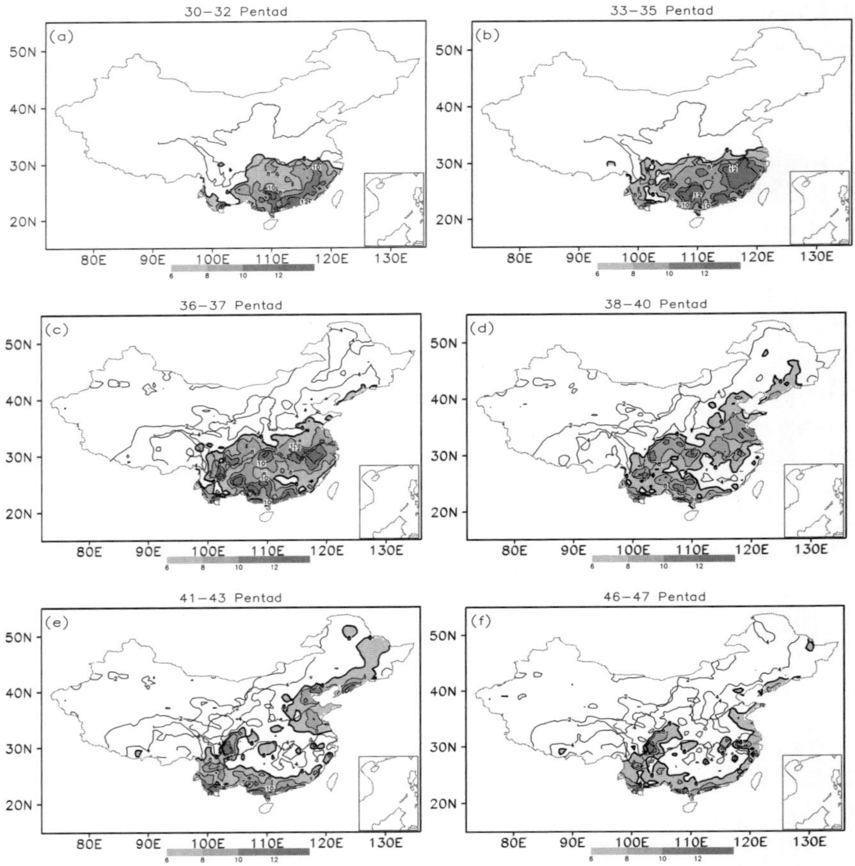

Fig. 3. Same as Fig. 2 except for the distribution of the rainfall in China average for 1958–2001.

Season over South China comes into the peak period and the heavy storms occur in South China frequently (Fig. 3(a)).

2. *EASM strengthen and jump northward to the mid-lower of Yangtze Valley* At the 33rd pentad, as the West Pacific subtropical high jump northward at first time, the southeasterly flows from the western subtropical high join the southwesterly flows from SCS at the mid-lower of Yangtze Valley and southwest of Japan (Fig. 2(b)) and the EASM's front jump northward to the mid-lower of Yangtze Valley and southwest of Japan. The front stays

at the mid-lower of Yangtze Valley steadily with less movement during the 33rd–35th pentad, Yangtze Valley begin Meiyu period (Fig. 3(b)).

3. *EASM advance northward to Yangtze–Huaihe Valley*

During the 36th–37th pentad, as West Pacific subtropical high moves northward further, the front of monsoon advance northward to the Yangtze–Huaihe Valley, Japan, Korea (Fig. 2(c)), these regions come into Huaihe Meiyu period (Fig. 3(c)), Baiu, Changma, respectively.

4. *EASM jump northward to North China and then reach the northernmost border*

During the 38th–40th pentad, as the West Pacific subtropical high appears the second time northward jump, the monsoon extend northward to North China further (Fig. 2(d)). The rain belt also moves to North China and the rainfall center stay at the lower of the Yellow River (Fig. 3(d)), but the precipitation in South China and central China is decreases rapidly and begin a dry period. During the 41st–42nd pentad, the West Pacific subtropical high extends northward continuously, and southwesterly flows move northeastward to Northeast China and the monsoon front reach the northernmost border of EASM (Fig. 2(e)). The precipitation of North China and Northeast China achieve maximum (Fig. 3(e)). In the 43rd pentad, the EASM's front keeps on stay at Northeast China, but intensity was a little weaker than that in the 41st–42nd pentad.

5. *EASM retreat southward rapidly*

During the 44th–45th pentad, the EASM's front begins to retreat southward and the southwesterly flows begin to weaken visibly. The intensity of precipitation over Northern China and Northeastern China also reduce rapidly. As the monsoon retreats southward quickly from 46th pentad (Fig. 2(f)), the EASM's front and the rain belt have moved southward to near 20°N and the summer rainfall in East China decrease rapidly at the 47th pentad (Fig. 3(f)).

The advance and retreat of EASM behave in a stepwise way, but not in a continuous way, it undergoes two abrupt northward shift at 33rd pentad and 38th pentad and three standing stages during the 30th–32nd pentad, 33rd–35th pentad and 38th–43rd pentad. The process that EASM retreat southward was much faster than when EASM moved northward. And the transfer of rain belt over East China is consistent with the advance and retreat of the EASM, that is, the EASM's front arrive at the region where the rainy season will begin, therefore, the early or late arrival of EASM's front, rapid or slow movement can directly influence the temporal and spatial distribution of the summer rainfall in East China.

3. The Abrupt Changes of EASM in Interdecadal Timescale

From the above discussion, we know that the EASM's movement influenced the rainfall in East China mainly during the 28th–45th pentad. As EOF technology develops especially in recent 10 years, it has been one of the most important climate diagnosis methods. Firstly, the time average EASM index (Im) (for 28th–45th pentad) is made every year and each data sample is subtracted from the average for 44 data sample in each grid, getting 44-year gridded abnormalities of EASM index from 1958 to 2001, then 44 abnormalities of EASM index are analyzed (1958–2001) in area (15°N–55°N, 105°E–150°E) by EOF method. Essentially, EOF made the evolution of physical fields divided into a few independent evolutive processes. Based on the method of North *et al.*,[16] the first three modes cleared the significant test. The variance contributions of three modes are 28.2%, 18.0%, 12.6%, respectively. In order to discuss the interdecadal characters of EASM's movement, the time coefficient of the second eigenvector exhibits three distinct interdecadal shifts in 1965, 1980, and 1994, respectively (Fig. 4(b)), the second mode is studied mainly in this paper. The second eigenvector present that the abnormal EASM's index is positive in north of the mid-lower of Yangtze Valley while it is negative in south of the mid-lower of Yangtze Valley (Fig. 4(a)). If the time coefficient is positive, indicating that the EASM is stronger (weaker) or stay longer (shorter) than the normal in north (south) of the mid-lower of Yangtze Valley, and vice versa. During the period of 1958–1964, 1965–1979, 1980–1993, and 1994–2001, the mean time

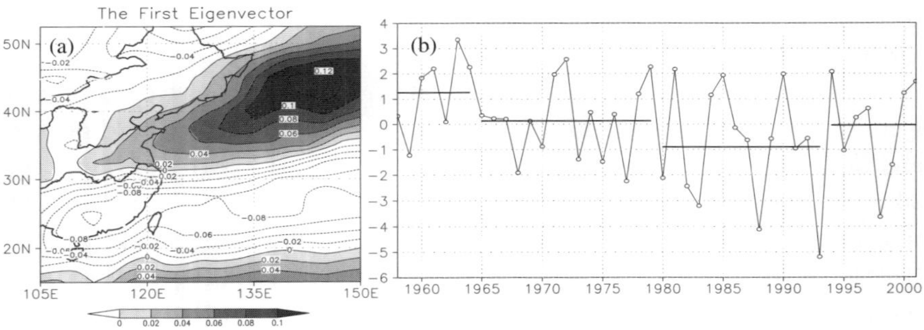

Fig. 4. (a) The second eigenvector. (b) The 1958–2001 curve for the time coefficient of the second eigenvector.

coefficient of the second mode are 1.26, 0.13, −0.9 and −0.04, respectively. This time coefficient indicate that the intensity of EASM is stronger and stay at North China and Northeast China longer prior to 1965, but EASM stayed at Yangtze River shorter. This situation changed greatly at 1965, especially after 1980, the mean time coefficient is −0.9 during the period of 1980–1993, and in this period EASM stayed longer and was stronger in south of Yangtze River while the time it stayed was shorter in North China. However, the situation changed again in 1994.

4. The Variability of EASM and the Spatial Distribution of Rainfall in East China

From Sec. 3, we know that the EASM's movement exhibits interdecadal abrupt shift in 1965, 1980, and 1994. We divided the EASM movement into four stages from 1958 to 2001 based on the time coefficient changes of the second eigenvector and give out the mean process of EASM's movement in each stage (Fig. 5).

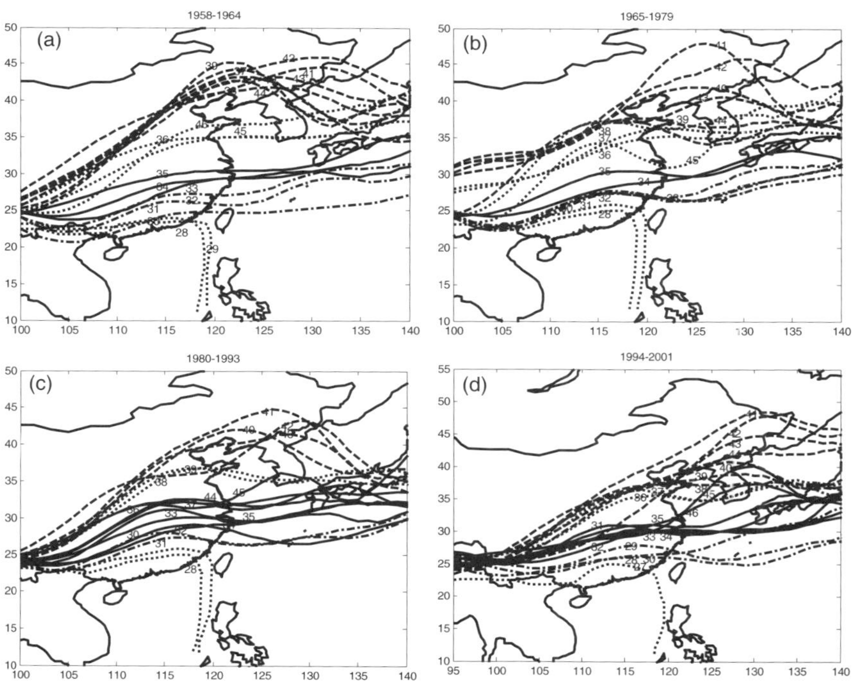

Fig. 5. The mean fronts of EASM for (a) 1958–1964, (b) 1965–1979, (c) 1980–1993, and (d) 1994–2001.

(I) *During the Period of 1958–1964*

Compare with the climatic field (average of 1958–2001), the Asian summer monsoon break out in SCS in the 28th pentad. The EASM's front is near normal during the 30th–32nd pentad (Fig. 5(a)), but it was south to normal obviously during 33rd–35th pentad while it is north to normal markedly during the 37th–40th pentad and during the 44th–46th pentad. EASM have advanced northward to North China in the 37th pentad and EASM retreat southward a little and it still stay at North China, and Northeast China in the 44th pentad. Moreover, it does not move southward to Yangtze River and South China staying at nearby North China during 45th and 46th pentad. But EASM retreat south of 20°N in the 47th pentad rapidly. In this period, the abnormal EASM's index is positive and abnormal south wind appears in north of Huaihe River (Fig. 7(a)), this is consistent with the EASM's movement, EASM is stronger and it stay longer in this region. Therefore, the abnormal movement of EASM directly caused the abnormal rainfall in East China, that is, EASM's front was south to normal obviously during the 33rd–35th pentad make the precipitation less above 15% in the mid-lower of Yangtze Valley and more precipitation in South China, while the front concentrate North China and Northeast China mainly during the 37th–45th pentad, bringing more above 15% precipitation to these regions. As a result, the abnormal EASM's movement brings more floods (drought) to South China, North China, and Northeast China (the mid-lower of Yangtze Valley). Using station rainfall data from 1881 to 1998, Kripalani and Kulkarni[17] pointed out that Indian summer monsoon have a abrupt change in 1963 and Indian summer monsoon became weaker after 1963, the change of Indian summer monsoon is consistent with the change of East Asian summer monsoon, so we can get a result that the whole Asian summer monsoon undergo an interdecadal abrupt change in the mid-1960s. They also revealed that the rainfall variation over North China is in-phase with South Asian rainfall, especially during 1930–1970. Therefore, the rainfall both over North Chain and India is much more than normal. In this period, the spatial distribution pattern of the precipitation anomaly percentage is "+, −, +, +" from South China, Yangtze River, North China to Northeast China (Fig. 6(a)). "+" refers to more rainfall than normal, and "−" to less rainfall.

(II) *During the Period of 1965–1979*

EASM break out in SCS at the 28th pentad, the EASM's front is north to normal a little during the 30th–31st pentad (Fig. 5(b)) However, it is south

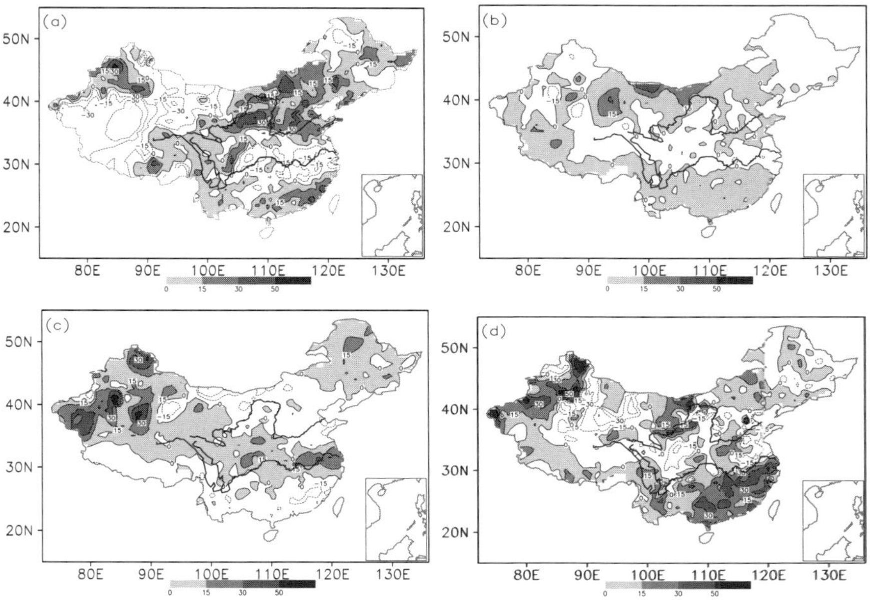

Fig. 6. Same as Fig. 5 except for the distribution of the rainfall percentage anomalies in China (average for 28th–45th pentad).

to normal obviously during the 33rd–34th pentad and during the 38th–39th pentad, staying at south of Yangtze River and near the lower of Yellow River during the 33rd–34th and the 38th–39th pentad, respectively. But it is north to normal obviously in 45th pentad with stay at the lower of Yellow River. EASM retreat south of 20°N and the rainfall in East China decreased remarkably in the 47th pentad. In this period, the EASM's movement presents the obvious regional characters, during 30th–34th pentad, the front mainly concentrate in South China and it stay at the lower of Yellow River during 36th–38th pentad and 45th pentad, and the abnormal EASM index was positive in above two regions (Fig. 7(b)) and southwesterly anomalies cover North China. But negative EASM's index anomalies and easterly anomalies cover the mid-lower of Yangtze Valley. The abnormal movement of EASM brought more precipitation to South China and North China, but it caused the persistent droughts in the Yangtze Valley (Fig. 6(b)). Though the precipitation in South China and North China was more than normal, it was less than previous period. In this period, the spatial distribution pattern of the precipitation anomaly percentage is "+, −, +, −" from South China to Northeast China (Fig. 7(b)).

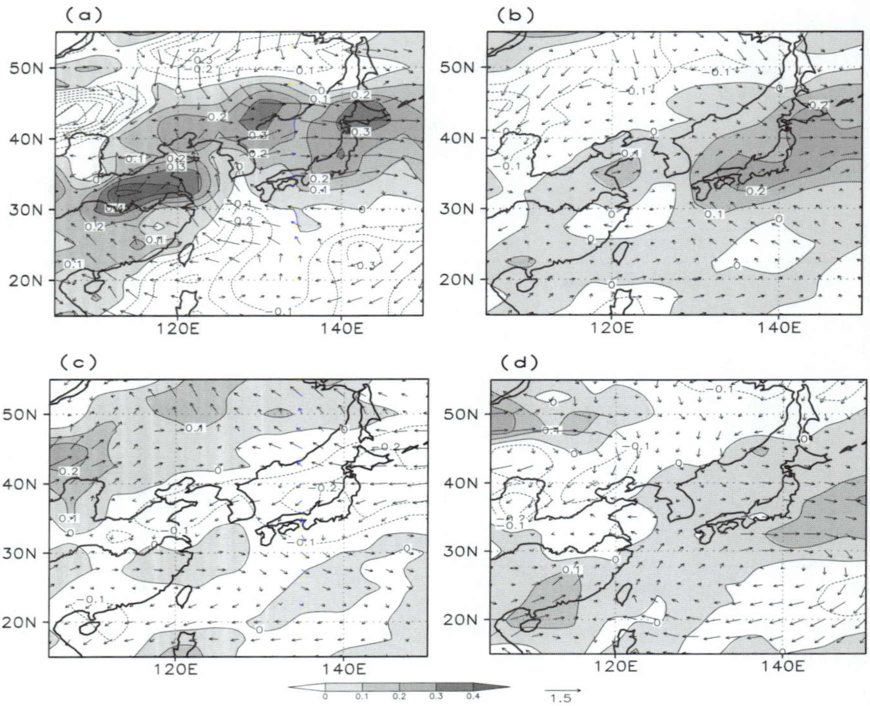

Fig. 7. Same as Fig. 5 except for the distribution of abnormal wind and EASM's index (average for 28th–45th pentad).

(III) *During the Period of 1980–1993*

EASM break out in SCS in the 28th pentad as normal, the EASM's front is north to normal in the 32nd pentad (Fig. 5(c)), but it was south to normal obviously during 35th–37th pentad, the front concentrate the mid-lower of Yangtze Valley mainly during 32nd–37th pentad. The front was also south to normal and stayed at the lower of Yellow River during 38th–39th pentad. The front stayed at North China and Northeast China steadily during 40th–41st pentad and 43rd pentad. The front was south to normal obviously in west of 120°E during the 42nd pentad and stayed at nearby Shandong provinces, and EASM retreat south of 20°N in 47th pentad. The positive EASM's index anomalies cover the mid-lower of Yangtze Valley while negative anomalies cover South China and North China. Westerly anomalies appear in Yangtze River and northerly flows control South China and North China (Fig. 7(c)). This indicates that EASM's front stayed longer than normal in Yangtze River Valley while it stayed shorter at South

China and North China. Compared with the previous stage, the abnormal movement of EASM causes the precipitation to increase markedly and brings much more floods to mid-lower Yangtze Valley, and the rainfall in Northeast China increase a little. However, the precipitation in South China and North China decrease obviously. In this period, the spatial distribution pattern of the precipitation anomaly percentage "−, +, −, +" from South China, Yangtze River, North China to Northeast China (Fig. 6(c)).

(IV) *During the Period of 1994–2001*

EASM break out in SCS in the 27th pentad early and the EASM's front is north to normal in 27th and the north edge of EASM has reached South China (Fig. 5(d)). The EASM's front was north to normal obviously during 28th–29th pentad and it stay at South China. And the front was also north to normal obviously during 31st–32nd pentad staying at the mid-lower of Yangtze Valley, but the front is south to normal in 35th pentad. Therefore, the front of EASM mainly concentrates the middle-lower of Yangtze River during 31st–35th pentad, bringing much more precipitation to this area. The front was also south to normal during 38th–40th pentad and it mainly stays at the lower of Yellow River during the 36th–40th pentad, causing more rainfall there. In the west of 115°E, the front is south to normal, but it is north to normal a little in east of 115°E and stay at Northeast China during 42nd–43rd pentad. But the front was north obviously in 44th pentad and 46th pentad, it is not retreat much and still it stay at Northeast China in the 44th pentad and it retreats to Yangtze River and not to South China as normal in the 46th pentad, EASM retreat south of 20°N in the 47th pentad. The positive EASM's index anomalies cover South China and the mid-lower of Yangtze Valley while negative anomalies cover North China. The Southwesterly anomalies appear in South China (Fig. 7(d)). The abnormal movement causes the precipitation increase markedly in South China and Yangtze River and the percentage anomalies of rainfall reaches above 30% in South China and the mid-lower of Yangtze Valley. Compare with the previous stage, the precipitation increases much more in South China and Yangtze River, especially in South China, but it decreases in North China continuously and this lead to the serious drought in North China. In this period, the spatial distribution pattern of the precipitation anomaly percentage is "+, +, −, +" from South China, Yangtze River, North China to Northeast China (Fig. 6(d)).

To summarize, as the EASM's movement exhibit three times interdecadal shifts in 1965, 1980, and 1994, respectively, the advance and retreat of EASM also exhibits interdecadal variability and this is the direct reason that cause the spatial distribution pattern of the rainfall in East China change. The abnormal rain belts move from north to south during 1958–2001. The East China underwent from the pattern of south-drought and northern-waterlog before 1979 to south-waterlog and north-drought.

5. Summary and Discussion

1. The advance and retreat of EASM behave in a stepwise way, but not in a continuous way. It undergo two abrupt northward jump at 33rd pentad and 38th pentad and three standing stages during the 30th–32nd pentad, 33rd–35th pentad, 36th–37th pentad, and 38th–43rd pentad in South China, the mid-lower Yangtze Valley, Huaihe Valley and North China and Northeast China, respectively.
2. The transfer of rain belt in East China is in consonance with the advance and retreat of EASM.
3. The movement of EASM experience three times interdecadal shifts in 1965, 1980, and 1994 respectively.
4. In the first stage (1958–1964), the front stays at South China mainly during the 30th–34th pentad, and it concentrates nearby North China and Northeast China during the 37th–46th pentad. Only in 35th pentad stays at the mid-lower of Yangtze River. The abnormal rain belt covers South China, North China and Northeast China, but the precipitation in Yangtze River is less than normal above 15%. During 1965–1979, the front concentrates South China and near the lower of Yellow River during 30th–34th pentad and 36th–38th pentad, respectively. The abnormal precipitation mainly covers South China and North China. During 1980–1993, the front of EASM stays at the middle and lower of Yangtze River during the 32nd–37th and 44th–45th pentad. The abnormal rain belt mainly concentrates Yangtze Valley. During 1994–2001, EASM break out earlier, the front concentrate South China during the 28th–30th pentad continuously and it stay at the mid-lower of Yangtze River steadily during the 31st–35th pentad and the 45th–46th pentad. The abnormal movement of EASM causes much more rainfall in South China, the mid-lower of Yangtze River. But it brought the serious drought in North China continuously.
5. The abnormal movement of EASM directly causes the shift of the spatial distribution pattern of rainfall in East China. During 1958–1964, the spatial distribution pattern of the precipitation anomaly percentage is "+, −, +, +" from South China, Yangtze River, North China to

Northeast China, but the pattern turns to "+, −, +, −" during 1965–1979, and it transit to "−, +, −, +" during 1980–1993, and then it become to "+ ,+, −, +". The precipitation in East China underwent from the pattern of south-drought and northern-waterlog before 1979 to south-waterlog and north-drought.

6. The precipitation increased in North China and Northeast China from 1994. Will the spatial distribution pattern of precipitation anomaly percentage in East China transit from the pattern of south-waterlog and north-drought to south-drought and northern-waterlog in the future? When will happen? We will make further study.
7. In this paper, we just give out the interdecadal variability of EASM's movement. How is the activity of EASM in the interdecadal timescales? What cause the interdecadal variability? These questions should be studied further.

Acknowledgments

This work was supported by National Natural Science Foundation of China (Grant No. 40775047) and Guangzhou Institute of Tropical and Oceanic Meteorology Science Foundation of China Meteorological Administration (Grant No. 200507).

References

1. L. Chen, Q. Zhu and H. Luo, *Monsoon in Asia China* (China Meteorological, Beijing, 1991), pp. 28–49.
2. Y. Ding, *Advanced Synoptic Meteorology* (China Meteorological, Beijing, 2004), pp. 212–249.
3. R. Huang, L. Zhou and W. Chen, The progresses of recent studies on the variabilities of the East Asian monsoon and their causes, *Adv. Atmos. Sci.* **20** (2003) 55–69.
4. Z. Zhang, A. Xie and R. Bai, Variability of East Asian summer monsoon and its association with rainfall trend over Songhuajiang–Nenjiang River Basin, *Meteorol. Sci. Technol.* **34** (2006) 542–546.
5. Q. Guo, J. Cai, X. Shao and W. Sha, Interdecadal variability of East-Asian summer monsoon and its impact on the climate of China, *Acta Geographica Sinica* **58** (2003) 569–576.
6. C.-G. Wu, H.-S. Liu and A. Xie, Interdecadal characteristics of the influence of northward shift and intensity of summer monsoon on rainfall over northern China in summer, *Plateau Meteorol.* **24** (2005) 656–665.
7. H. Hu and W. Qian, Define the boundary belt for EASM, *Progress Natl. Sci.* **17** (2007) 57–65.

8. M. Kwon, J.-G. Jhun, B. Wang *et al.*, Decadal change in relationship between East Asian and WNP summer monsoon, *Geophys. Res. Lett.* **32** (2005) L16709.
9. S. Wang, D. Gong, J. Ye and Z. Chen, Seasonal precipitation series of eastern China since 1880 and the variability, *Acta Geographica Sinica* **55** (2000) 281–292.
10. Z. Zhao, *The Flood-Drought and General Circulation in China* (China Meteorological, Beijing, 1992), pp. 1–10.
11. R. Huang, Y. Xu and L. Zhou, The interdecadal variation of summer precipitation in China and the drought trend in North China, *Plateau Meteorology* **18** (1999) 465–476.
12. R. Huang and R. Cai, Interdecaldal variations of drought and flooding disasters in China and their association with the East Asian climate system, *Chin. J. Atmos. Sci.* **30** (2006) 730–743.
13. J. Xu and M. Wei, The climate change features in the last 100 years, *J. Capital Normal Univ.* **27** (2006) 79–82.
14. X. Shi and X. Xu, Characters of the interdecadal transition of climate in summer and winter in China, *Chin. Sci. Bull.* **15** (2006) 2075–2084.
15. X. Lu and X. Zhang *et al.*, The interdecadal abrupt change of the onset and advance of East-Asian summer monsoon, *Acta Oceanologica Sinica*, in press.
16. G. R. North, T. Bell, R. Cahalan *et al.*, Sampling errors in the estimation of empirical orthogonal function, *Mon. Wea.* **110** (1982) 699–706.
17. R. H. Kripalani and A. Kulkarni, Monsoon rainfall variations and teleconnections over South and East Asia, *Int. J. Climatol.* **21** (2001) 603–616.

Advances in Geosciences
Vol. 10: Atmospheric Science (2007)
Eds. J. H. Oh and G. P. Singh

FORMATION OF TROPICAL CYCLONE CONCENTRIC EYEWALLS BY WAVE–MEAN FLOW INTERACTIONS

JIAYI PENG
International Pacific Research Center, University of Hawaii,
1680 East West Road, Honolulu, Hawaii, USA

TIM LI
Department of Meteorology and International Pacific Research Center,
School of Ocean and Earth Science and Technology,
University of Hawaii, Honolulu, Hawaii, USA

MELINDA S. PENG
Naval Research Laboratory, Monterey, California, USA

The role of two-way interactions between a symmetric core vortex and an asymmetric disturbance in generating tropical cyclone (TC) concentric eyewalls is examined in a nonlinear barotropic model. The results show that when an asymmetric perturbation is placed outside of the radius of maximum wind, an asymmetric disturbance develops in the inner region, resulting in a weakening of the symmetric flow *in situ*, while the symmetric tangential wind gains energy from the asymmetric perturbations in the outer region. This process leads double peaks in the symmetric tangential wind profile. Further diagnosis reveals that the distinctive evolution features in the inner and outer regions are determined by the asymmetric up- (down-) shear tilting structure and so-induced symmetry-to-asymmetry (asymmetry-to-symmetry) energy transfer. There exists an optimal radius location for the initial perturbation to generate most efficiently a double-peak structure in the symmetric tangential wind profile.

1. Introduction

Concentric eyewalls have been observed in the life cycle of strong tropical cyclones (TC). Willoughby *et al.*[12] identified a double eyewall structure for hurricane Gilbert with the inner eyewall in the radius of 8–20 km and the outer eyewall between 55 and 100 km. A more detailed analysis of Gilbert[1] showed that the primary eyewall appeared first. During a weakening stage of the storm, the outer eyewall formed. Later on, the outer eyewall strengthened and contracted while the inner eyewall weakened.

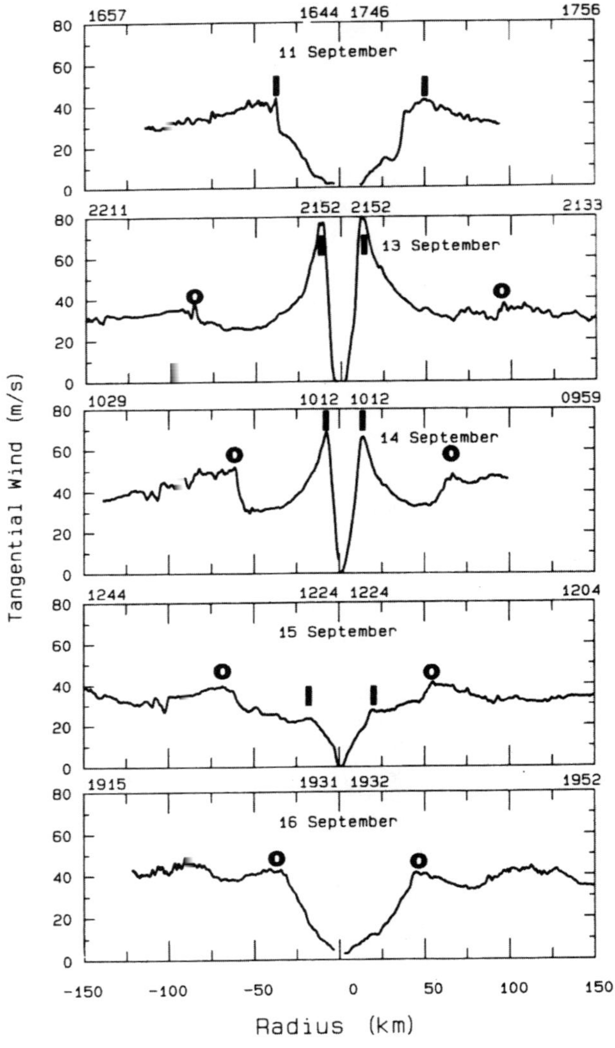

Fig. 1. Flight-level tangential wind speed from south to north traverses through the center of Hurricane Gilbert. Both 'I' and 'O' denote the location of the inner and outer eyewall wind maximum, respectively. Times at the beginning and end of each radial pass are plotted at the top of the panels (refer to 1).

Finally, the outer eyewall replaced the inner eyewall and completed an eyewall replacement cycle (Fig. 1).

Several studies have devoted to understand mechanisms through which concentric eyewalls form. Willoughby *et al.*[10] and Willoughby[11] suggested a symmetric instability that might contribute to the formation of the outer eyewall. They, however, could not develop a causal relation between the location of the outer eyewall and the instability. Montgomery and Kallenbach[3] implied that the TC concentric eyewalls could result from radially propagating linear vortex Rossby waves that are dynamically constrained near a critical radius. Since the development and propagation of the vortex Rossby waves are attributed to the TC basic state radial vorticity gradient, the vortex Rossby waves are confined near the radius of the maximum wind (RMW). Nong and Emanuel[4] studied the formation of the concentric eyewalls in an axisymmetric model. Their simulations showed that the secondary eyewall might result from a finite-amplitude WISHE instability, triggered by external forcing.

Black and Willoughby[1] noted that the outer eyewall formed during the TC weakening stage (Fig. 1). Shapiro and Willoughby[7] and Willoughby *et al.*[9] used a symmetric model (hereafter SW model) to diagnose the secondary circulation induced by a point heat source in balanced, axisymmetric vortices. For a heat source near RMW, a maximum of the tangential wind tendency lay just inside of RMW, so that the maximum wind propagated inward in response to the heating, which provided a plausible physical explanation for the contraction of the outer wind maximum. However, their simulations did not reproduce a double-peak structure. The fact that an outer eyewall forms during TC weakening stage suggests that a rapid decrease of convective heating may play a role in the formation of double eyewalls. Peng *et al.*[5] examined this idea by introducing a negative heat source in a simple TC model.

Theoretically, concentric eyewalls may be formed as two cyclonic vortices with different sizes and intensities interact without merging into a monopole.[2] The criteria for the concentric eyewall formation is that (1) the core vortex must be at least six times stronger in vorticity than the neighboring weaker vortex, (2) the neighboring vortex is larger in size than the core vortex, and (3) a separation distance is within three to four times of the core vortex radius. Note that in this scenario, a symmetric positive vortcity belt has been given initially. The interaction between the two vortices just redistributes the vorticity of the outer vortex. In this study, we present a different, wave–mean flow interaction scenario. We examine a new concentric eyewall formation scenario in which a core vortex interacts with an asymmetric perturbation that has a wave-like structure and zero symmetric vorticity component in the outer region. We will examine how the symmetric flow gains energy from the asymmetric perturbation in the outer region, and how the second peak of the symmetric tangential wind is induced.

The outline of this paper is as follows. A brief description of the model and the experimental design is given in Sec. 2. Results from the nonlinear simulations are discussed in Sec. 3. Finally, a summary is given in Sec. 4.

2. Model Description

2.1. *The model*

To study the interaction between the core vortex and asymmetric perturbation, in particular to study how the asymmetry influences the symmetric flow, we construct a nonlinear barotropic model. The governing equations in a non-dimensional form on an f plane are given as the following:

$$\frac{\partial u}{\partial t} + u\frac{\partial u}{\partial x} + v\frac{\partial u}{\partial y} - v = -\frac{\partial \phi}{\partial x}, \tag{2.1a}$$

$$\frac{\partial v}{\partial t} + u\frac{\partial v}{\partial x} + v\frac{\partial v}{\partial y} + u = -\frac{\partial \phi}{\partial y}, \tag{2.1b}$$

$$-2J(u,v) - \zeta = -\nabla^2\phi, \tag{2.1c}$$

where $J(u,v) = \frac{\partial u}{\partial x}\frac{\partial v}{\partial y} - \frac{\partial v}{\partial x}\frac{\partial u}{\partial y}$ and $\zeta = \frac{\partial v}{\partial x} - \frac{\partial u}{\partial y}$, u and v are the horizontal wind velocity components; ϕ, the geopotential height, and ζ is the vorticity. The Coriolis parameter is $f = 5 \times 10^{-5}\,\text{s}^{-1}$, and the characteristic value for time is $T = 1/f = 2 \times 10^4\,\text{s}$. A non-dimensional time $t = 0.18$ corresponds to 1 h. The characteristic values for the velocity and horizontal length scales are $C = 50\,\text{m/s}$ and $L = CT = 1000\,\text{km}$, and the Rossby number equals 1 for the vortex.

The numerical solution technique employed is the fourth-order Runge–Kutta scheme with a time increment of 0.002. The Matsuno scheme[8] is applied to calculate the advection terms, and the second-order centered difference is used for the approximation of other space derivatives. A second-order diffusion is applied every 0.18 time with the non-dimensional coefficient being 1.4×10^{-6} to ensure the numerical stability. The model covers a 2×2 (2000 km × 2000 km) area with a grid resolution of 0.002 in both x and y directions. The lateral boundary condition is radiative. All simulations are carried out for 24 h. Most of the results shown are up to 12 h during which the major axisymmetrization process occurs.

2.2. *Experiment design*

A core vortex is specified initially, which has a hurricane-like tangential wind profile (see Fig. 2) defined as follows:

$$V(r/R_{\max}) = V_{\max}\frac{2(r/R_{\max})}{1+(r/R_{\max})^2}, \tag{2.2}$$

where the maximum tangential wind $V_{\max} = 0.5(25\,\mathrm{m/s})$, and the radius of maximum wind $R_{\max} = 0.1(100\,\mathrm{km})$. The vorticity of the core vortex has a maximum at the center of the vortex (as seen in Fig. 1(c)) and decreases

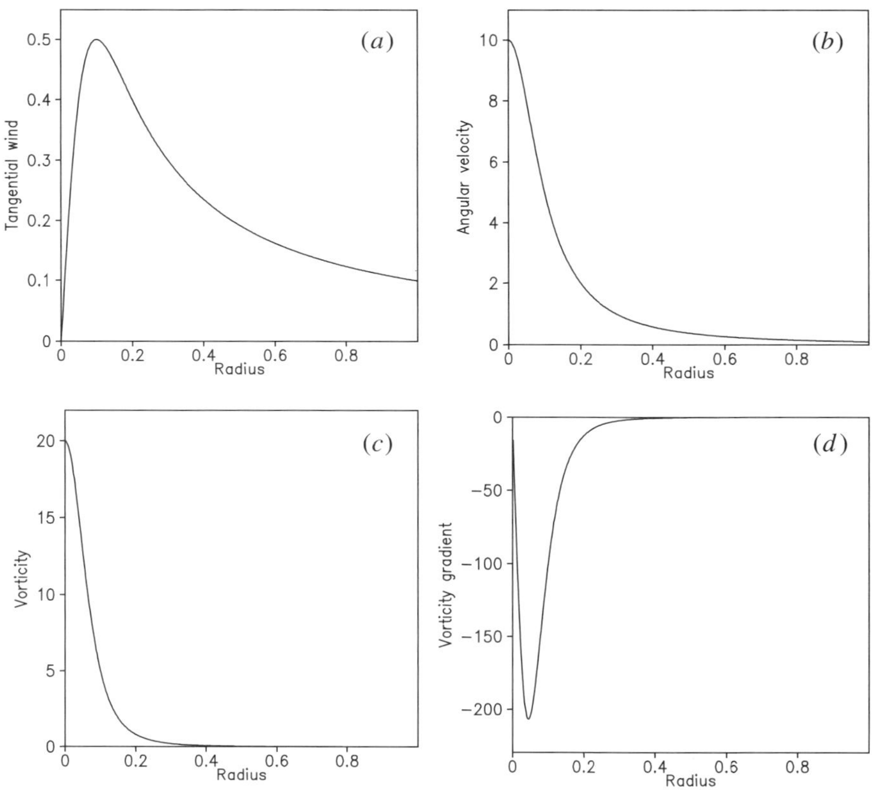

Fig. 2. Radial profiles of the hurricane-like vortex for nondimensional (a) tangential wind, (b) angular velocity, (c) vorticity, and (d) vorticity gradient. To obtain tangential wind in ms^{-1}, multiply by 50. To obtain angular velocity or vorticity in s^{-1}, multiply by 5×10^{-5}. To obtain vorticity gradient in $\mathrm{m}^{-1}\mathrm{s}^{-1}$, multiply by 5×10^{-11}. To obtain radial displacement in km, multiply by 1000.

monotonically with radius, with a maximum vorticity gradient located at $r = 0.044$ (Fig. 2(d)).

Different from the vortex–vortex interaction scenario in Kuo *et al.*,[2] we focus on the interaction between the asymmetric disturbances and symmetric core vortex flows. The initial asymmetry specified contains either a wavenumber 2 or a wavenumber 3 structure in the azimuthal direction. (To avoid the movement of the core vortex, a wavenumber-one asymmetry is not considered.) The initial asymmetry is prescribed by a vorticity perturbation with the following expression:

$$\varsigma' = 5 \exp\left[-\frac{1}{2}\left(\frac{r - R_{\mathrm{p}}}{\sigma}\right)^2\right] \cos(k\lambda), \tag{2.3}$$

where r is the radial distance; λ the azimuthal angle; k, the azimuthal wavenumber ($k = 2$ or $k = 3$) and the radial scale (or size) of the asymmetry $\sigma = 0.025$. The radial parameter R_{p} controls the position of the initial asymmetry.

To investigate how the initial asymmetry position might affect the formation of the second peak of the symmetric tangential wind, five experiments have been designed for the wavenumber 2 perturbations. In the first experiment the initial perturbation is placed at the radius of 0.2 ($R_{\mathrm{p}} = 0.2$, hereafter denoted as T20, see Fig. 3(a)). In the second

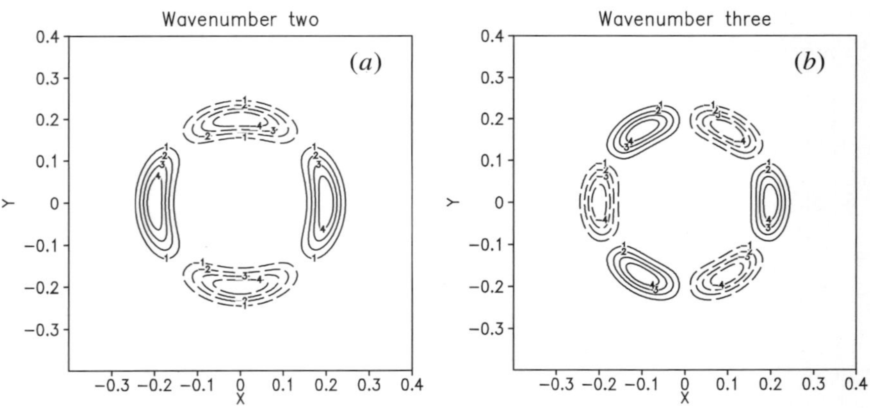

Fig. 3. The initial non-dimensional barotropic asymmetric vorticity with the maximum center located at the radius of 0.2 for (a) wavenumber 2 case T20 and (b) wavenumber 3 case H20. The contour interval is 1. To obtain vorticity in s^{-1}, multiply by 5×10^{-5}. To obtain radial displacement in km, multiply by 1000. Only the inner 400 km × 400 km model domain is shown.

experiment the initial asymmetry is placed at the radius of 0.25 ($R_{\mathrm{p}} = 0.25$, hereafter denoted as T25); the third one at the radius of 0.3 ($R_{\mathrm{p}} = 0.3$, denoted as T30); the fourth at the radius of 0.1 ($R_{\mathrm{p}} = 0.1$, denoted as T10); and the fifth at the radius of 0.15 ($R_{\mathrm{p}} = 0.15$, denoted as T15). The similar five sensitivity experiments with the wavenumber 3 disturbances are denoted as H20 ($R_{\mathrm{p}} = 0.2$), H25 ($R_{\mathrm{p}} = 0.25$), H30 ($R_{\mathrm{p}} = 0.3$), H10 ($R_{\mathrm{p}} = 0.1$), and H15 ($R_{\mathrm{p}} = 0.15$) respectively (Fig. 3(b)).

2.3. *Diagnosis method*

The diagnosis of the model output is carried out in a cylindrical coordinate system centered at the vortex center. Each model variable is decomposed into a symmetric and an asymmetric component, e.g. $u = \bar{u} + u'$, $v = \bar{v} + v'$, with a bar denoting the symmetric component and a prime the departure from the symmetric field.

The diagnostics for the energy budget is made with the following symmetric kinetic energy [KE, $\bar{K} = \frac{1}{2}(\bar{u}^2 + \bar{v}^2)$] equation:

$$\frac{\partial \bar{K}}{\partial t} = -\frac{\partial (r\bar{u}\bar{K})}{r\partial r} - \bar{u}\frac{\partial (r\overline{u'^2})}{r\partial r} - \bar{v}\frac{\partial (\overline{u'v'})}{\partial r} + \bar{u}\frac{\overline{v'^2}}{r} - \frac{2\bar{v}}{r}\overline{u'v'} - \bar{u}\frac{\partial \bar{\phi}}{\partial r}, \qquad (2.4)$$

where the first term on the right-hand side of (2.4) is the flux divergence of $\bar{K}$ by the symmetric radial flow, the sum of the second, third, fourth, and fifth terms represents the time change rate of symmetric KE due to wave–wave interactions, and the sixth term is the energy conversion from symmetric potential energy to symmetric kinetic energy. Note that the second-to-fifth terms on the right-hand side involve the interaction among the asymmetric perturbations and they are directly related to energy transfer between the asymmetry and the symmetry.

A Fourier analysis for vorticity is based on the following formula:

$$\zeta(r, \lambda, t) = \zeta_0(r, t) + \sum_{k=1}^{N} [\zeta_{kc}(r, t)\cos(k\lambda) + \zeta_{ks}(r, t)\sin(k\lambda)], \qquad (2.5)$$

where k is the azimuthal wavenumber, $\zeta_0(r, t)$, $\zeta_{kc}(r, t)$, and $\zeta_{ks}(r, t)$ are Fourier spectrum coefficients. The wavenumber spectrum of the first eight components are used to calculate the asymmetric component, and the Fourier asymmetric vorticity amplitude is defined by the following formula:

$$A_k(r, t) = \sqrt{\zeta_{kc}(r, t)^2 + \zeta_{ks}(r, t)^2}.$$

3. Results

We first diagnose the result from the wavenumber 2 perturbation simulations. Figure 4 shows the time evolution of the simulated symmetric tangential wind profile at a 3-h interval. As one can clearly see, a double-peak wind profile appears at hour 6. After that, the outer maximum continues to grow, while the inner peak experiences an oscillation in amplitude. For instance, at hour 9, there are two peaks in the symmetric tangential wind profile, with the inner one retaining bigger amplitude; at hour 12, the outer one has stronger amplitude.

A key question related to this double eyewall formation is how the outer wind peak is established. A notable feature is that the symmetric tangential wind in the outer region ($r > 0.15$) continues to grow while the wind in the inner region oscillates after initial rapid decay. These distinctive evolution features between the outer and inner regions are closely related to the energy transfer between the symmetric flow and the asymmetric perturbation, as shown in Fig. 5.

The diagnosis of the energy exchange between the symmetric and asymmetric components (the second-to-fifth terms on right-hand side of (2.4)) shows that outside of $r = 0.15$ there is always a positive energy

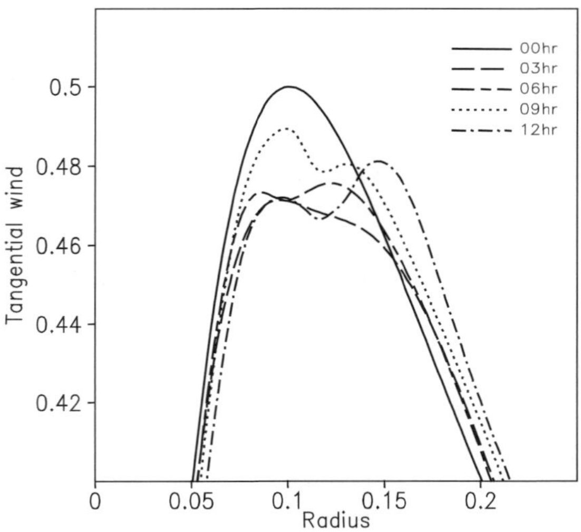

Fig. 4. The evolution of non-dimensional symmetric tangential wind profiles for case T20. To obtain tangential wind in ms^{-1}, multiply by 50. To obtain radial displacement in km, multiply by 1000.

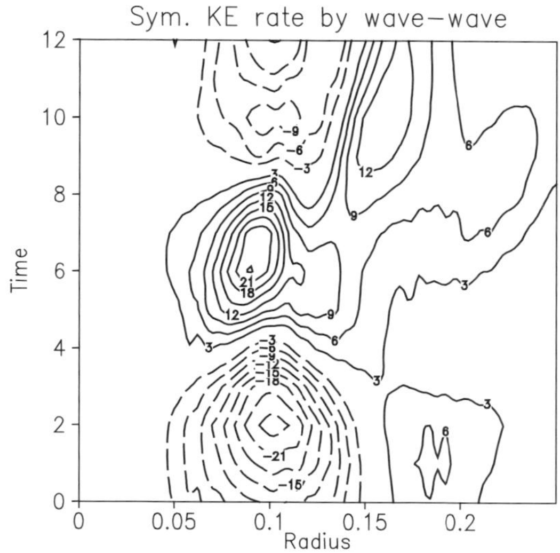

Fig. 5. The time–radius cross-section of the asymmetry-to-symmetric kinetic energy transfer rate (unit: $1.25 \times 10^{-4}\,\mathrm{m^2s^{-3}}$) in association with the wave–wave interactions in case T20. To obtain radial displacement in km, multiply by 1000. The time unit is hour.

transfer from the asymmetric perturbation to the symmetric flow, whereas inside of this radius there is oscillatory behavior in the energy transfer, that is, the symmetric flow gains energy from the asymmetry during hours 4–9 but loses energy into the asymmetry during hours 0–4 and 9–12. This is consistent with the time tendency of the symmetric tangential wind near RMW.

To understand the cause of the distinctive energy transfer behavior, we examine the asymmetric perturbation structure and its evolution characteristics. Figure 6 shows the time evolution of amplitude of the asymmetric perturbation. Note that in this numerical experiment (T20) the initial wavenumber 2 asymmetry is placed at the radius of 0.2. After time integration, a strong asymmetry is generated within the first four hours inside the radius of 0.1 where the absolute value of the symmetric vorticity gradient is the largest (Fig. 2(d)). The asymmetry amplitude in the outer region ($r > 0.15$), however, decreases gradually.

The horizontal pattern of the asymmetric perturbation reveals that the asymmetric vorticity field exhibits distinctive patterns during the different development stages. For example, at hour 1, the phase line connecting this newly generated asymmetry inside of RMW and the original outer

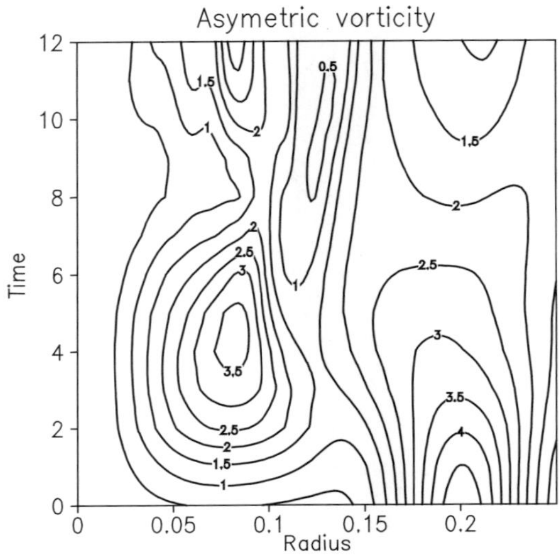

Fig. 6. The time–radius cross-section of the asymmetric vorticity amplitude (unit: $5 \times 10^{-5}\,\mathrm{s}^{-1}$) for case T20. To obtain radial displacement in km, multiply by 1000. The time unit is hour.

asymmetry shows an up-shear tilt (Fig. 7(a)) with respect to the rotation angular velocity of the core vortex (Fig. 2(b)). Because of this up-shear tilt, the symmetric flows transfer their energy to the asymmetric perturbations near $r = 0.1$ before hour 4 (Fig. 5), resulting in the weakening of the symmetric core vortex at hour 3 (Fig. 4). Because the symmetric angular velocity advects the inner asymmetry at a much faster rotation rate (see the angular velocity profile in Fig. 2(b)) than the outer asymmetry, the asymmetric vorticity shifts its phase to a down-shear tilt in the period of hours 4–9 (refer to Fig. 7(b)), so that the energy is transferred back to the symmetric flows (Fig. 5) and the asymmetric vorticity amplitude decreases during the period (Fig. 6). Thus, the symmetric tangential wind near the radius of 0.1 grows at the expense of the weakening of the asymmetry from hour 6 to hour 9 (Fig. 4). A new up-shear-tilting inner asymmetry is induced again after hour 9 (Figs. 6 and 7(c)). As a result, the symmetric flows transfer their energy to the asymmetric perturbations during hours 9–12, while the tangential wind at the inner core region weakens (Fig. 4).

In contrast, the symmetric flows always gain energy from the asymmetric disturbances in the outer region ($r > 0.15$) due to the steady down-shear phase tilt of the asymmetric disturbances (Fig. 7). This causes

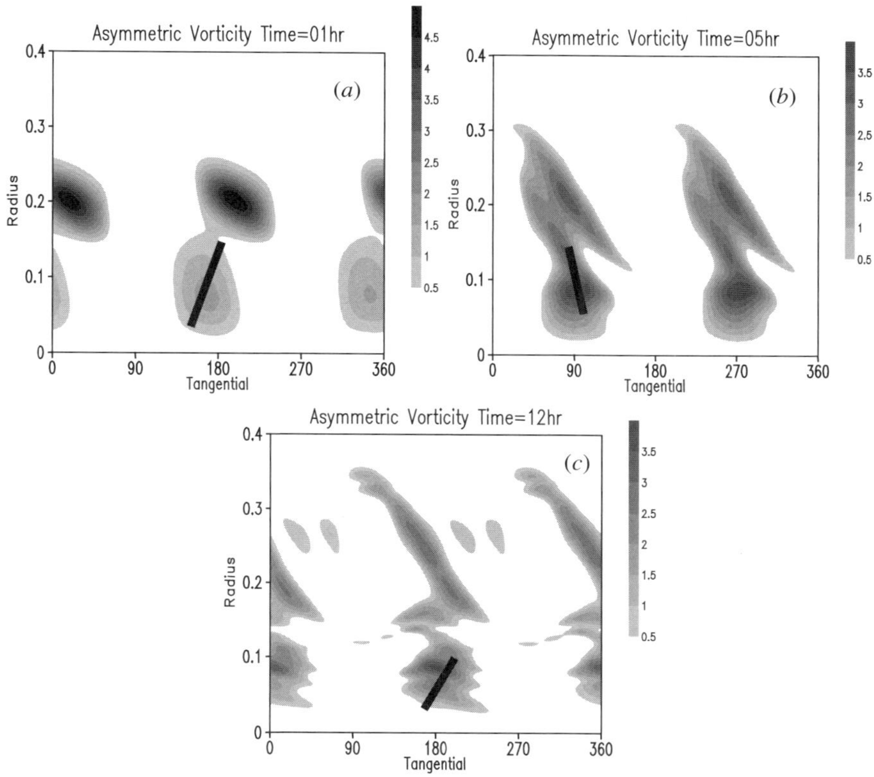

Fig. 7. The non-dimensional asymmetric vorticity pattern at time (a) 01 h, (b) 05 h and (c) 12 h for case T20. To obtain vorticity in s^{-1}, multiply by 5×10^{-5}. To obtain radial displacement in km, multiply by 1000.

the continuous intensification of the tangential wind in the outer region, leading to the formation of the second peak in the symmetric tangential wind profile.

To examine whether the aforementioned wind evolution characteristics change with different initial perturbations, we conduct a set of parallel experiments in which an initial wavenumber 3 asymmetry is introduced. Figure 8 shows the symmetric tangential wind evolution in case H20, where the initial asymmetric perturbation is placed at the radius of 0.2. Compared to case T20, a weaker asymmetry is generated near the radius of 0.1 (Fig. 9(a)). The comparison of symmetric kinetic energy change rates between the wavenumber 2 (T20) and wavenumber 3 (H20) perturbation

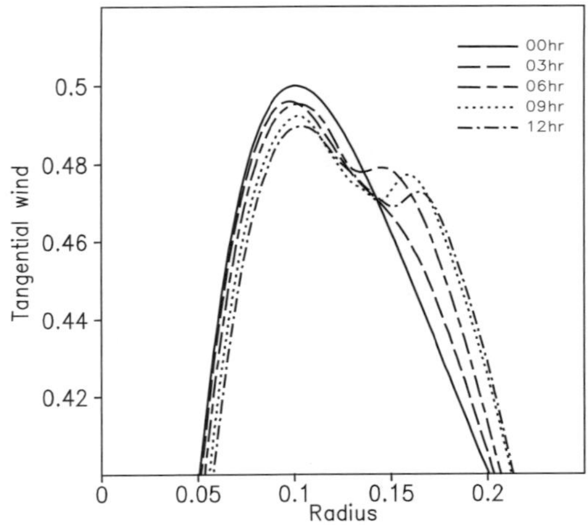

Fig. 8. The evolution of non-dimensional symmetric tangential wind profiles for wavenumber 3 initial disturbances in case H20. To obtain tangential wind in ms^{-1}, multiply by 50. To obtain radial displacement in km, multiply by 1000.

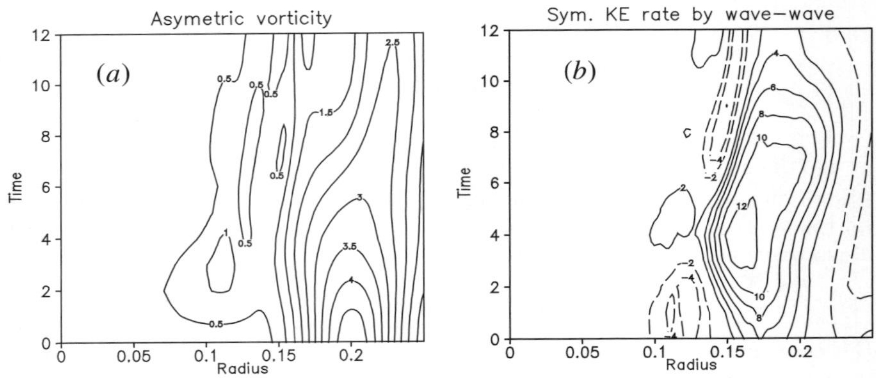

Fig. 9. The time–radius cross-section of (a) the asymmetric vorticity amplitude (unit: $5 \times 10^{-5}\,s^{-1}$) and (b) the asymmetry-to-symmetry kinetic energy transfer rate (unit: $1.25 \times 10^{-4}\,m^2\,s^{-3}$) for wavenumber 3 initial disturbances in case H20. To obtain radial displacement in km, multiply by 1000. The time unit is hour.

experiments indicate that the energy exchange between the asymmetric and symmetric flows is weaker in the wavenumber 3 case. Nevertheless, a weak oscillation of the energy transfer is still present near the radius of 0.1, while in the outer region ($r > 0.15$) the wavenumber 3 initial disturbance can transfer more energy into the symmetric flows than the corresponding

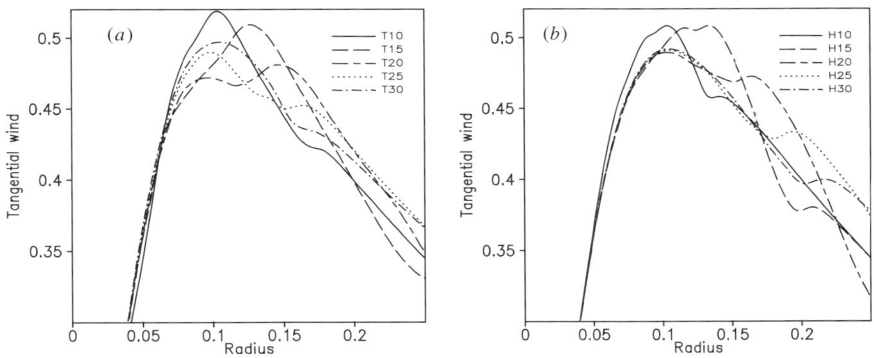

Fig. 10. The non-dimensional symmetric tangential wind profiles at 12 h for (a) wavenumber 2 and (b) wavenumber 3 initial disturbances. To obtain tangential wind in ms^{-1}, multiply by 50. To obtain radial displacement in km, multiply by 1000.

wavenumber 2 disturbance (Fig. 9(b)). It is again the down-shear tilt of the asymmetric perturbation and so induced asymmetry-to-symmetry energy transfer in the outer region that generate the second peak in the symmetric tangential wind profile (Fig. 8).

For the same-structure initial perturbation, is there a preferred radius location for the double eyewall formation? Our sensitivity experiments with the same wavenumber 2 or 3 initial perturbation but with different radial locations show that indeed there exists such an optimal radius. When the perturbation is placed more outward (i.e. T25, T30, H25, and H30) compared to the T20 and H20 experiments, the second peak in the symmetric tangential wind profile becomes weaker (Fig. 10), which means that the symmetric flows gain less energy from the asymmetric perturbations. On the other hand, when the initial asymmetry is placed more inward in T15 and T10 (H15 and H10), there is no obvious second-peak in the tangential wind profile. Thus, the sensitivity experiments above point out an optimal location near $r = 0.2$ (i.e. twice of RMW), where the initial asymmetry may generate the most significant double peaks in the symmetric tangential wind profile (Fig. 10).

4. Summary

The role of two-way interactions between a symmetric core vortex and an asymmetric disturbance in generating TC concentric eyewalls is examined in a nonlinear barotropic model. The results show that when an asymmetric perturbation is placed at twice of RMW, an asymmetric disturbance develops in the inner core region, resulting in a weakening of the symmetric tangential wind. However, the symmetric flow gains energy from the asymmetric perturbations in the outer region, which induces the second peak of the symmetric tangential wind. This process is robust for both the wavenumber 2 and 3 perturbations, pointing out a new wave–mean flow interaction scenario for the double eyewall formation.

The numerical simulations illustrate that two distinctive symmetry–asymmetry interaction regimes in the inner and outer regions, respectively. While the symmetric tangential wind exhibits an oscillatory evolution in the inner region, it grows steadily in the outer region. This distinctive evolution feature is closely related to the asymmetric vorticity pattern, its up- or down-shear tilt, and so-induced symmetry-to-asymmetry or asymmetry-to-symmetry energy transfer.

Sensitivity numerical experiments indicate that there exists an optimal radius location (approximately near twice of the radius of the maximum wind) where the initial asymmetric disturbance may generate the most significant double-peak structure in the tangential wind profile. The optimal radius exists in both the wavenumber 2 and 3 experiments.

In the current study, a simple nonlinear barotropic model is used. Further studies with more sophisticated models are needed to validate the wave–mean flow interaction processes.

Acknowledgments

This work was supported by ONR grants N000140710145 and N000140210532, NRL subcontract N00173-06-1-G031, and National Natural Science Foundation of China under Grants 40205009 and 40333025. The International Pacific Research Center is partially sponsored by the Japan Agency for Marine-Earth Science and Technology (JAMSTEC). This is SOEST publication number 1234 and IPRC publication number 123.

References

1. M. L. Black and H. E. Willoughby, The concentric eyewall cycle of hurricane Gilbert, *Mon. Wea. Rev.* **120** (1992) 947–957.
2. H.-C. Kuo, L.-Y. Lin, C.-P. Chang and R. T. Williams, The formation of concentric vorticity structures in typhoons, *J. Atmos. Sci.* **61** (2004) 2722–2734.

3. M. T. Montgomery and R. J. Kallenbach, A theory for vortex Rossby-waves and its application to spriral bands and intensity changes in hurricanes, *Quart. J. Roy. Meteorol. Soc.* **123** (1997) 434–565.
4. S. Nong and K. A. Emanuel, A numerical study of the genesis of concentric eyewalls in hurricane, *Quart. J. Roy. Meteorol. Soc.* **129** (2003) 3323–3338.
5. J. Y. Peng, J. D. Jou, M. S. Peng, J. Fang and R. S. Wu, The formation of concentric eyewalls with heat sink in a simple tropical cyclone model, *Terrest. Atmos. Ocean. Sci.* **17** (2006) 111–128.
6. W. H. Schubert, J. J. Hack, P. L. Silva Dias and S. R. Fulton, Geostrophic adjustment in an axisymmetric vortex, *J. Atmos. Sci.* **37** (1980) 1461–4483.
7. L. J. Shapiro and H. E. Wiloughby, The response of balanced hurricanes to local sources of heat and momentum, *J. Atmos. Sci.* **39** (1982) 373–894.
8. T. L. Shen, Y. X. Tian, X. Z. Ge, W. S. Lu and D. H. Chen, *Numerical Weather Prediction* (Beijing Meteorology Press, 2003), pp. 145 (in Chinese).
9. H. E. Willoughby, J. A. Clos and M. G. Shoreibah, Concentric eye walls, secondary wind maxima, and the evolution of the hurricane vortex, *J. Atmos. Sci.* **39** (1982) 394–511.
10. H. E. Willoughby, H. L. Jin, St. J. Lord and J. M. Piotrowicz, Hurricane structure and evolution as simulated by an axisymmetric nonhydrostatic numerical model, *J. Atmos. Sci.* **41** (1984) 1161–9186.
11. H. E. Willoughby, The dynamics of the tropical cyclone core, *Aust. Met. Mag.* **36** (1988) 181–391.
12. H. E. Willoughby, J. M. Masters and C. W. Landsea, A record minimum sea level pressure observed in Hurricane Gilbert, *Mon. Wea. Rev.* **117** (1989) 2824–2828.

Advances in Geosciences
Vol. 10: Atmospheric Science (2007)
Eds. J. H. Oh and G. P. Singh

TROPICAL CIRCULATION INDICES AND PERFORMANCES OF INDIAN SUMMER MONSOON RAINFALL

G. P. SINGH*, JAI-HO OH†, S. N. PANDEY* and R. BHATLA*
*Department of Geophysics, Banaras Hindu University, Varanasi 221 005, India

†Department of Environment and Atmospheric Sciences, Pukyong National University, Busan, South Korea

The interannual relationships between the summer monsoon rainfall over all India (AIR), northwest India (NWR), and peninsular India (PIR), and seven different tropical circulation indices (TCIs) (based on mean sea level pressure) over five selected tropical stations, three over the Indian Ocean namely Agalega (A), Cocos Island (C), Il-Nouvelle (I), and two over the land stations namely New Delhi (N) and Malacol (M) have been examined for 30 years' period (1953–1982). The names of the indices are (i) TCI (A–M), (ii) TCI (A–N), (iii) TCI (A–C), (iv) TCI (C–M), (v) TCI (C–N), (vi) TCI (I–N), and (vii) TCI (I–M). The results indicate that significant strong and inverse relationships exist between (a) TCI (C–M) of concurrent August and AIR, NWR, and PIR, (b) TCI (C–N) of antecedent February and AIR, NWR, and PIR and (c) TCI (I–N) of antecedent May shows significant and direct association with AIR and PIR. Stability analysis of different TCIs shows that TCI (C–M) of concurrent August and TCI (C–N) of antecedent February show consistently significant relationship over the successive 25 years' period.

1. Introduction

An extended low pressure area is observed over the northwest India, Pakistan, and Saudi Arabia, and even up to the northeastern part of Africa during the northern summer. The pressure gradient created between the Mascarene Hgh over the Indian Ocean to this extended low generally develops during the spring season and persists during the monsoon season also. It controls the cross-equatorial flow. The strength of the summer monsoon circulation which brings moist air from the Indian Ocean and causes rainfall over India and its neighborhood depends on this gradient. Several authors[1–3] have found that transequatorial pressure gradient measured by Southern Oscillation Index (SOI) is not a useful predictor of the Indian summer monsoon, because it shows significant correlations with the concurrent and the following year's SOI rather than the antecedent SOI. For this reason, the present authors created other pressure indices herein after called Tropical Circulation Indices (TCIs) between pairs of stations, as a measure of pressure gradients in the tropics and trans-Indian Ocean, and examined potential links between Indian monsoon rainfall and TCIs with the hope that some of them might lead to a better correlation than SOIs.

Seven tropical circulation indices (TCIs) using monthly mean values of mean sea level pressure (MSLP) at five selected tropical stations have been computed. We then made a detailed study of the lead/lag teleconnections between Indian summer monsoon rainfall and the TCIs mentioned above, using the data for the period 1953–1982, which may find useful application in long-range prediction of the summer monsoon rainfall over India. Earlier, Hastenrath,[4] Parthasarthy *et al.*,[5,6] and Singh and Chattopadhyay[7] examined the relationships between the Indian monsoon rainfall and regional/global circulation parameters.

2. Data and Procedure of Analysis

As already mentioned, the TCIs which measures the cross-equatorial flows were computed using the mean sea level pressure (MSLP) over the following five tropical stations. These are Il Nouvelle (37°08′S, 77°05′E), Agalega (10°06′S, 56°08′E), Cocos Island (12°02′S, 96°08′E), New Delhi (28°38′N, 77°12′E) and Malacol Island (13°02′N, 32°07′E). The first three stations present the strength of the Mascarene High over the Indian Ocean, while the last two land stations lie over the equatorial trough of low pressure during the summer monsoon and spring seasons. MSLP values over these five stations were obtained from the Monthly Climatic Data for the World, NOAA, USA over the 30 years' period 1953–1982. From these five stations, seven TCIs were computed for different months using MSLP for the following combination of stations.

(i) Agalega–Malacol and the corresponding TCI being referred to as TCI (A–M).
(ii) Agalega–New Delhi and the corresponding TCI as TCI (A–N).
(iii) Agalega–Cocos Island and the corresponding TCI being referred to as TCI (A–C).

(iv) Cocos Island–Malacol and the corresponding TCI being referred to as TCI (C–M).
(v) Cocos Island–New Delhi and the corresponding TCI being referred to as TCI (C–N).
(vi) Il Nouvelle–New Delhi and the corresponding TCI being referred to as TCI (I–N).
(vii) Il Nouvelle–Malacol and the corresponding TCI being referred to as TCI (I–M).

The methods of computation of TCIs are as follows. In order to study the low-frequency oscillation, annual oscillation was filtered from MSLP data. This has been done by subtracting the 30 years' (1953–1982) mean monthly values from the respective individual months. This monthly anomaly time series was standardized using the standard deviation of the respective anomaly time series. Then, difference is further standardized by the standard deviation of the difference time series. This method is used for all the above seven groups of stations for all the 12 months from January to December to obtain the respective monthly TCI series. From these TCIs, four standard seasonal series centered at January, April, July, and October are also made. These are used in this study. The rainfall series for the summer monsoon months (June to September) over India (AIR) and two of its subregions viz., the Northwest India (NWR) and the Peninsular India (PIR) have been prepared by Chattopadhyay and Bhatla.[2] They have used area-weighted subdivisional rainfall in India during the period 1901–1990. This data series from 1953 to 1982 was used in the present study.

To establish the relationships between the all India summer monsoon rainfall and TCIs, lead/lag and contemporaneous correlation coefficients (CCs) were computed among them for several months/seasons prior to the summer monsoon season and several months/seasons after the summer monsoon season. Consistencies of the significant CCs were also tested using sliding windows of different widths. The statistical significance of the correlations was tested using the conventional two-tailed t-test.

3. Results and Discussion

To investigate the temporal degree of association among the seven different TCIs and Indian summer monsoon rainfall anomalies viz., AIR, NWR, and PIR, concurrent and lagged correlations (CCs) were computed. Thus, AIR, NWR, and PIR seasonal monsoon rainfall anomalies have been correlated with various monthly/seasonal TCIs series. These months and seasons are (i) the months from January to May and two seasons DJF (December, January, and February) and MAM (March, April, and May), antecedent to summer monsoon season, (ii) the months from June to September and a season JJA (June, July, and August), concurrent to the summer monsoon, and (iii) the months October, November, December and two seasons SON+ (September, October, and November), and DJF+ (December, January, and February), the succeeding periods with respect to the summer monsoon season.

The analysis shows several significant correlations on which our discussion will be restricted. Table 1 shows the CCs between Indian summer monsoon rainfall and TCI series for the period 1953–1982 along with their levels of significance.

As far as concurrent relationships are considered, Table 1 shows significant negative CC of −0.37 (at 5% level) between TCI (A–M) JJA and NWR. In addition, significant negative CCs are also obtained between concurrent August TCI (A–M) and AIR, NWR, and PIR (CC = −0.51, −0.51, and −0.48, all at 1% level). September TCI (A–M) and PIR show

Table 1. Correlation coefficients (CCs × 100) between Indian summer monsoon rainfall and monthly/seasonal TCIs.

Months/Season	AIR	NWR	PIR	Months/Season	AIR	NWR	PIR
TCI(A–M) Aug(0)	$-51^{(1)}$	$-51^{(1)}$	$-48^{(1)}$	TCI(C–N) Feb(−)	$-50^{(1)}$	$-42^{(2)}$	$-38^{(5)}$
TCI(A–N) Sep(0)	−33	−24	$-42^{(2)}$	TCI(C–N) Apr(−)	$-39^{(5)}$	−17	$-38^{(5)}$
TCI(A–M) JJA(0)	−22	$-37^{(5)}$	−15	TCI(C–N) Jun(0)	$-42^{(2)}$	$-37^{(5)}$	$-44^{(2)}$
TCI(A–N) May(−)	$36^{(5)}$	33	20	TCI(C–N) Jul(0)	$-37^{(5)}$	$-38^{(5)}$	−28
TCI(A–N) Jul(0)	$-47^{(1)}$	$-47^{(1)}$	$-42^{(2)}$	TCI(C–N) JJA(0)	$-44^{(2)}$	$-43^{(2)}$	$-38^{(5)}$
TCI(A–N) JJA(0)	−32	$-42^{(2)}$	−33	TCI(C–N) Nov(+)	$-56^{(1)}$	−34	$-48^{(1)}$
TCI(A–N) DJF(+)	$-46^{(1)}$	−32	$-52^{(1)}$	TCI(C–N) Dec(+)	$-42^{(2)}$	−26	$-37^{(5)}$
TCI(A–C) Feb(−)	$36^{(5)}$	$46^{(1)}$	23	TCI(C–N) SON(+)	$-38^{(5)}$	−22	−27
TCI(A–C) Oct(+)	$37^{(5)}$	$36^{(5)}$	18	TCI(C–N) DJF(+)	$-41^{(5)}$	−15	$-39^{(5)}$
TCI(A–C) Nov(+)	$49^{(1)}$	28	$40^{(5)}$	TCI(I–N) Jan(−)	−31	$-39^{(5)}$	−27
TCI(A–C) SON(+)	$41^{(5)}$	23	23	TCI(I–N) May(−)	$46^{(1)}$	31	$39^{(5)}$
TCI(C–M) Aug(0)	$-70^{(.1)}$	$-66^{(.1)}$	$-58^{(.1)}$	TCI(I–N) Jul(0)	$-43^{(2)}$	−23	$-39^{(5)}$
TCI(C–M) JJA(0)	$-51^{(1)}$	$-49^{(1)}$	$-41^{(5)}$	TCI(I–N) Sep(0)	29	14	$36^{(5)}$
TCI(C–M) Oct(+)	$-46^{(1)}$	$-39^{(5)}$	−35	TCI(I–N) JJA(0)	$-39^{(5)}$	−33	$-37^{(5)}$
TCI(C–M) Nov(+)	$-41^{(5)}$	−24	−34	TCI(I–N) DJF(+)	$-61^{(.1)}$	$-61^{(.1)}$	$-57^{(.1)}$
TCI(C–M) SON(+)	$-54^{(1)}$	$-41^{(5)}$	$-43^{(2)}$	TCI(I–M) Jan(−)	$-37^{(5)}$	$-50^{(1)}$	−30
				TCI(I–M) DJF(−)	−16	$-40^{(5)}$	−16
				TCI(I–M) Aug(0)	$-51^{(1)}$	$-49^{(1)}$	$-41^{(5)}$
				TCI(I–M) JJA(0)	$-36^{(5)}$	−28	−28
				TCI(I–M) DJF(+)	$-61^{(.1)}$	$-64^{(.1)}$	$-57^{(.1)}$

Notes: $^{(.1)}$, $^{(1)}$, $^{(2)}$ and $^{(5)}$ stand for 0.1%, 1%, 2%, and 5% significant level; (−), (0), (+) stand for antecedent, concurrent, and succeeding periods.

significant negative CC of −0.42 (at 2% level). But Table 1 does not show any significant relationship between antecedent and succeeding TCI (A–M) and AIR, NWR, and PIR.

For TCI (A–N), Table 1 shows significant negative CC of −0.42 (2% level) between concurrent TCI (A–N) JJA and NWR. The CCs between July TCI (A–N) and AIR, NWR, and PIR are −0.47 and −0.47 (both at 1% level) and −0.42 (2% level), respectively, while succeeding relationships show significant negative CCs of −0.46 (1% level) and −0.52 (1% level) between TCI (A–N) of DJF and AIR and PIR, respectively. However, the antecedent relationship shows significant positive CC of 0.36 (5% level) between TCI (A–N) of May and AIR only.

Concurrent TCI (A–C) does not show any significant relationship with AIR, NWR and PIR (Table 1). However, succeeding October, November, and SON show significant positive CCs of 0.37 (5% level), 0.49 (1% level), and 0.41 (5% level) with AIR; succeeding October TCI (A–C) with NWR (CC = 0.36 at 5% level), and succeeding November TCI (A–C) with PIR (CC = 0.40 at 5% level). As far as antecedent relationships are concerned, Table 1 shows that only February TCI (A–C) have significant positive CCs of 0.36 (5% level) with AIR and 0.46 (1% level) with NWR. Thus, it appears that February TCI (A–C) has the potential to be used as a predictor for Indian summer monsoon rainfall all over India and the northwest India.

For TCI (C–M), Table 1 shows the strong and inverse CCs of −0.70, −0.66, and −0.58 (all are significant at 0.1% level) between concurrent August TCI (C–M) and AIR, NWR, and PIR. Significant negative CCs are also obtained between concurrent JJA TCI (C–M) and AIR, NWR, and PIR (CC = −0.51 at 1% level), −0.49 (at 1% level), and −0.41 (at 5% level), respectively. Figure 1 shows the year to year relationships between the anomalies of TCI (C–M) of concurrent August, and JJA and AIR showing the inverse association between them, while significant negative CCs are obtained between succeeding October TCI (C–M) and AIR and NWR (CC = −0.46 at 1% level, −0.39 at 5% level), respectively. November TCI (C–M) with AIR shows negative CC of −0.41 at 5% level and succeeding SON with AIR, NWR, and PIR (CC = −0.54 at 1% level with AIR, −0.41 at 5% level with NWR, and −0.43 at 2% level with PIR). However, there is no significant relationship between antecedent TCI (C–M) and AIR, NWR, and PIR.

Considering TCI (C–N), Table 1 shows significant negative CCs of −0.44, −0.43, and −0.38 at 2% and 5% levels between concurrent JJA and AIR, NWR and PIR, respectively. In addition, significant negative

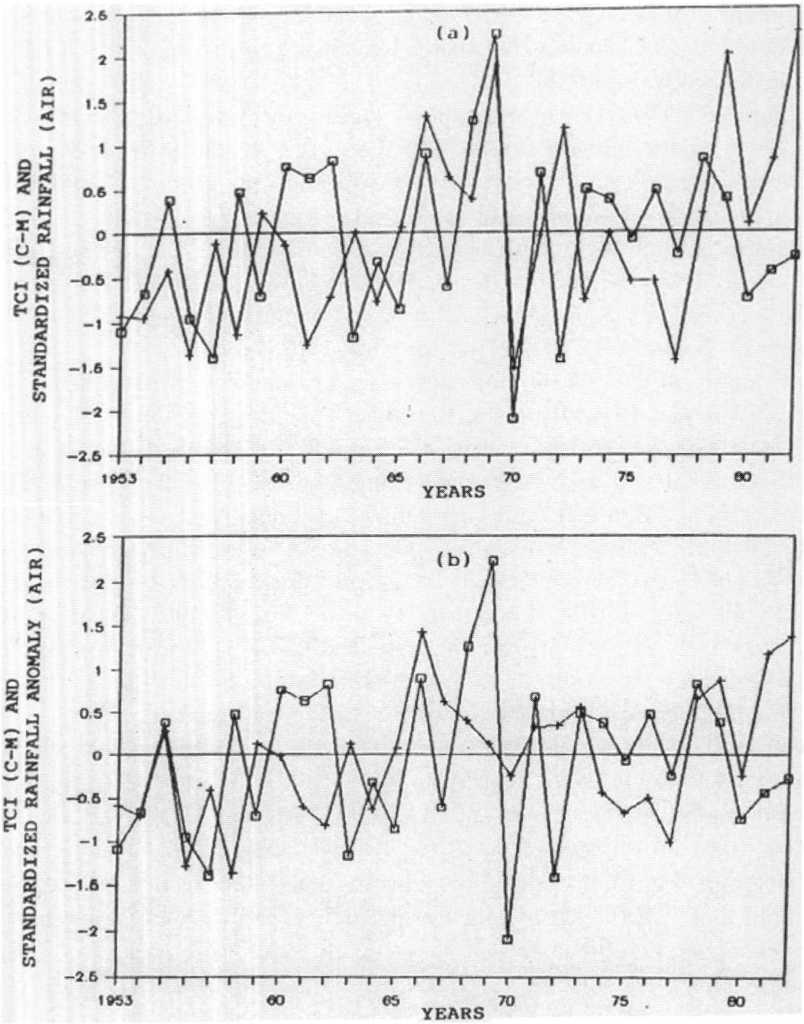

Fig. 1. Time series of standardized rainfall anomaly AIR (marked by square) and TCI (C–M) of concurrent (a) August (marked by plus) and (b) JJA (marked by plus).

CCs are also obtained between concurrent June and July TCI (C–N) and AIR (CC $= -0.42$ at 2% level and -0.37 at 5% level), and with NWR (CC $= -0.37$ and -0.38), both significant at 5% level. Concurrent June TCI (C–N) and PIR show CC of -0.44 at 2% level, while succeeding November, December, SON, and DJF show significant and negative CCs of -0.56

(1% level), −0.42 (2% level), −0.38 (5% level), and −0.41 (5% level) with AIR and PIR. Succeeding November, December, and DJF TCI (C–N) show negative CCs of −0.48 (1% level), −0.37 (5% level), and −0.39 (5% level). As far as antecedent relationships are concerned, TCI (C–N) of February shows strong and significant negative CCs of −0.50 (1% level) with AIR, −0.42 (2% level) with NWR, and −0.38 (5% level) with PIR, while April TCI (C–N) shows significant negative CCs of −0.39 and −0.38 both at 5% level with AIR and PIR. Thus, February TCI (C–N) appears to be a potential predictor for AIR, NWR, and PIR.

For TCI (I–N), it is found that concurrent JJA shows significant negative CCs of −0.39 at 5% level with AIR and −0.37 (5% level) with PIR. In addition, significant negative CCs are also obtained between concurrent July TCI (I–N) and AIR (CC = −0.43 at 2% level) and PIR (CC = −0.39 at 5% level) and September TCI (I–N) with PIR (CC = −0.36 at 5% level). Succeeding DJF TCI (I–N) shows significant negative CCs of −0.61 (0.1% level) with AIR and NWR and −0.57 (0.1% level) with PIR. For antecedent relationship, Table 1 shows significant negative CCs of −0.39 (5% level) with January TCI (I–N) and NWR. The antecedent May TCI (I–N) shows significant CCs of 0.46 at 1% level and 0.39 at 5% level with AIR and PIR.

From the above analysis, it appears that May TCI (I–N) has some potential as a predictor for long-range forecasting of Indian summer monsoon rainfall over AIR and PIR. The strong and significant inverse CCs between TCI (I–N) of succeeding DJF suggest that a strong (weak) summer monsoon rainfall activity is followed by unusually high (low) pressure over northwest India and unusually low (high) pressure over Mascarene High during the following winter months.

Considering TCI (I–M), Table 1 shows significant negative CC of −0.36 (5% level) between concurrent TCI (I–M) JJA and AIR. TCI (I–M) of August shows significant negative CCs of −0.51 (1% level) with AIR, −0.49 (1% level) with NWR, and −0.41 (5% level) with PIR. Succeeding DJF shows negative CCs of −0.61 (0.1% level), −0.64 (0.1% level), and −0.57 (0.1% level) with AIR, NWR, and PIR, respectively. Significant relationship (CC = −0.40 at 5% level) is also found between antecedent TCI (I–M) DJF and NWR. In addition, significant negative CCs are also obtained between antecedent TCI (I–M) of January and AIR (CC = −0.37 at 5% level) and NWR (CC = −0.50 at 1% level).

To examine the consistency of significant TCI relationships, the CCs of different months for the sliding 25-year periods have been recomputed during the period 1953–1982. Stability analysis has been done only for those TCIs which display strong correlation (mostly for antecedent period). Stability analysis of antecedent February TCI (A–C) shows significant correlation over all the different periods of study with PIR, except for the first 25-year period viz., 1953–1982. It is interesting to note that the correlation is not significant with PIR during the full period, but it becomes significant during different periods of study except the earliest period. Hence, it can be used as a predictor for PIR.

Considering concurrent TCI (C–M) of August, the analysis shows that the correlation is highly significant over various periods of study from 1953 to 1982 for AIR, NWR, and PIR. For antecedent TCI (C–N) of February, correlation remains significant for AIR and PIR over different sliding periods of study. For TCI (C–N) of antecedent April, correlation is significant during 1953–1977 and 1954–1978 for AIR and NWR, and in 1955–1979 for AIR only. While TCI (I–N) of antecedent May shows that the correlation is significant over all the periods of study for AIR only and for NWR, correlation is significant during the period 1953–1977 only. However, for antecedent TCI (I–M) of January, no significant correlation is found during different sliding periods (although, it shows significant correlation during the full period of study). So it may not be a good predictor.

From the stability analysis of different TCIs, it appears that among all the above TCIs, TCI (C–N) of antecedent February shows the strongest negative CCs with the following all India and regional peninsular Indian summer monsoon rainfall, and it can be used as a potential for long-range forecasting of summer monsoon over AIR and PIR.

4. Conclusions

The statistical analysis between Indian summer monsoon rainfall and different Tropical Circulation Indices (TCIs) shows the following important results: Indian summer monsoon rainfall is strongly and inversely related to the concurrent TCI (C–M) of August. It means monsoon has very strong physical relationship with TCI (C–M). TCI (C–N) of antecedent February is strongly related to AIR and thus could be used as a predictor for all Indian summer monsoon rainfall. TCI (I–N) of antecedent May has significant association with all Indian summer monsoon rainfall. An examination of stability analysis shows that TCI (A–C) of previous February and TCI (I–N) of previous May are good predictors for PIR and AIR, respectively. TCI (C–N) of previous February is a good predictor for AIR and PIR.

Acknowledgments

The authors wish to express their sincere thanks to ADGM(R), India Meteorological Department, Pune for providing the necessary rainfall data. The authors would also like to thank the NOAA, United States for providing the Monthly Climatic Data for the World, on tape containing the mean sea level pressure data. This work was funded by the Korea Meteorological Administration Research and Development Program under Grant CATER 2006-11011. G. P. Singh would also like to acknowledge CATER for supporting his visit to the Pukyong National University, Busan, South Korea.

References

1. B. Parthasarathy and G. B. Pant, *Tellus* **36A** (1984) 269.
2. J. Chattopadhyay and R. Bhatla, *PAGEOPH* **141** (1993) 177.
3. J. Chattopadhyay and R. Bhatla, *Mausam* **47** (1996) 59.
4. S. Hastenrath, *J. Climate Appl. Meteorol.* **26** (1987) 847.
5. B. Parthasarathy, H. F. Diaz and J. K. Eischeid, *J. Geo. Res.* **93**(D5) (1988) 5341.
6. B. Parthasarathy, N. A. Sontakke, A. A. Munot and D. R. Kothawale, *Mausam* **41**(2) (1990) 301.
7. G. P. Singh and J. Chattopadhyay, *Mausam* **49**(4) (1998) 443.

Advances in Geosciences
Vol. 10: Atmospheric Science (2007)
Eds. J. H. Oh and G. P. Singh

THE TROPICAL PACIFIC–INDIAN OCEAN TEMPERATURE ANOMALY MODE AND ITS IMPACT ON ASIAN CLIMATES

YANG HUI*,‡ and LI CHONGYIN*,†

**State Key Laboratory of Numerical Modeling for Atmospheric Sciences and Geophysical Fluid Dynamics, Institute of Atmospheric Physics, Chinese Academy of Sciences, Beijing 100029, China*

‡*yanghui@mail.iap.ac.cn*

†*Meteorological College, PLA University of Science and Technology, Nanjing 211101, China*

The Indian Ocean temperature anomaly is very closely related to the Pacific Ocean temperature anomaly through the Walker circulation and the Indonesian through flow. So only the El Niño/Southern Oscillation (ENSO) in the Pacific cannot entirely explain the influence of sea surface temperature anomaly (SSTA) on climate variation. In this paper, the tropical Pacific–Indian Ocean temperature anomaly mode (PIM) is presented from pattern and feature of SSTA in both the Indian Ocean and the Pacific. Further, the features of PIM and ENSO mode and their influences on the climate in China and the rainfall in India are compared. The observation and sensitivity experiments show that presenting PIM and studying its influence are very important for short-range climate prediction. Furthermore, the characteristics of the Asian climates related to the Pacific–Indian Ocean temperature anomaly mode are investigated.

1. Introduction

The impacts of anomalous sea temperature on general circulation and climate have attracted researchers' attention. The strong anomalous sea surface temperature in the equatorial eastern Pacific El Niño Southern Oscillation (ENSO) event especially causes serious flood or drought disasters in many regions and countries around the globe and is an important subject of many studies for many years in the world.[1–4] Although ENSO is the strongest signal in the interannual climate variation, it is not the only cause of the anomalous climate. For example, the summer precipitations were more than normal in some regions of India[5,6] and eastern China[7] in some El Niño years. On the other side, the SSTA (sea surface temperature anomaly) in the Indian Ocean has been paid attention for a long time.[8] Moreover, the Indian Ocean temperature dipole was discovered a few years ago.[9,10] This work discussed the important influences of the dipole including the influence on the climate in China.[11–14] Recent study has also shown that the impact of the Indian Ocean dipole mode has more impact on the East Asian Monsoon (in particular over Korea–Japan) than over South Asia.[34] Generally, the Indian Ocean temperature dipole was significantly related to the ENSO mode in the Pacific.[15] And the equatorial western Pacific and the eastern Indian Ocean are all called warm pool.[16] The Indian through flow[17,18] and anomalous atmospheric zonal-vertical (Walker) circulation[15] are the main link pass for relating the SSTA in Indian Ocean to that in the Pacific. Since ENSO is the strongest signal in the interannual variation, generally, some climate phenomena including SSTA in the middle-high latitude ocean at all lag SSTA in the eastern equatorial Pacific.[19,20] But observation research has revealed that the correlation between the Pacific Ocean dipole (ENSO) and the Indian Ocean dipole is largely simultaneous with no significant lag.[15] The equatorial ocean SST change (response) occurring almost simultaneously is attributed mainly to the close interaction of the two anomalous atmospheric Walker circulations. On that account, the tropical Pacific–Indian Ocean temperature anomaly mode has clear physical meaning: the SSTA pattern and change in both the equatorial Pacific and the Indian Ocean are related to the two anomalous Walker circulations, that is, the two anomalous Walker circulations result in the SSTA patterns in the two Oceans, respectively. Apparently, those can be regarded as dipoles. In order to reveal the influence rule of the tropical ocean SSTA on climate, the tropical Pacific–Indian Ocean temperature anomaly mode (PIM) will be presented through analyzing its SSTA patterns and features. And the influences of every SSTA patterns on Asian climates are researched.

The data used are monthly SST data of the Hadley Centre in UK on a $1.0^\circ * 1.0^\circ$ grid from 1900 to 1999. The monthly rainfall and temperature data at 160 stations of China compiled by the China Meteorological Administration and the global land monthly precipitation data (PREC/L) on $2.5^\circ \times 2.5^\circ$ grids (1948–2001) are also used.

2. Tropical Pacific–Indian Ocean Temperature Anomaly Mode

El Niño and La Niña are defined by the SSTA in the equatorial eastern Pacific. Actually, when there is the positive (negative) SSTA in the eastern equatorial Pacific, the negative (positive) SSTA in the western equatorial Pacific occurs. On the other hand, the so-called Indian Ocean dipole is defined by the difference of the SSTA in the western equatorial Indian Ocean from that in the eastern equatorial Indian Ocean, indicating zonal heat contrast of the Indian Ocean SSTA. Although the name dipole is used, it actually does not define the mathematic meaning (SSTA distribution with positive (negative) in the west and opposite in the east).[15] Considering the close relationship between the Pacific ENSO mode and the Indian Ocean dipole, the index of the PIM can be defined as the respectively normalized east–west differences of the equatorial areas in the two oceans. As to the SSTA, the SSTA of ENSO is stronger than that in the equatorial Indian Ocean because of the bigger Pacific basin. However, as to the influence of the SSTA on East Asia, a series of numerical experiments clearly indicate that the effect of SSTA forcing of the Indian Ocean is stronger than that of the eastern equatorial Pacific.[21–23] So we will define the composite index based on the normalized dipoles respectively in the Pacific and the Indian Ocean. In this way, the normalized dipoles are comparable and the composite index has mathematical and physical base, so does its change. The composite index I_{com} is defined as

$$\begin{aligned} I_{\text{com}} &= \nabla T_i + \nabla T_{\text{p}}, \\ \nabla T_i &= T_1 - T_2, \\ \nabla T_{\text{p}} &= T_3 - T_4. \end{aligned}$$

The symbols T_1, T_2, T_3, and T_4 are denoted, respectively, by the averaged SSTA in the region (50°E–65°E, 5°S–10°N), (85°E–100°E, 10°S–5°N), (130°W–80°W, 5°S–5°N) and (140°E–160°E, 5°S–10°N). The symbols ∇T_i and ∇T_{p} are normalized.

Figure 1(c) gives the time series of I_{com} (tropical Pacific–Indian Ocean temperature anomaly mode index) and SSTA in Niño$_{3.4}$ (also called ENSO index). It can be found that although I_{com} is closely related to ENSO index, the difference in them is large. The importing Indian Ocean SSTA mostly makes the I_{com} stronger than the (normalized) ENSO mode, and partly

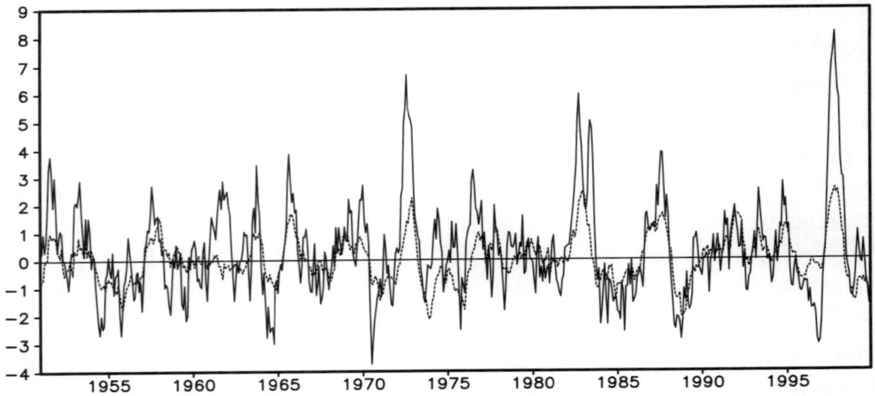

Fig. 1. Time series of the tropical Pacific–Indian Ocean temperature anomaly mode index (solid line), and SSTA in Niño$_{3.4}$ (dash line).

changes the phase of ENSO index. This indicates that studying composite mode and its index I_{com} are very important.

In order to further explain the difference between the composite mode and the pure Pacific ENSO mode, the two types of positive phase and quasi-normal phase are compared. Based on the tropical Pacific–Indian Ocean temperature anomaly mode index, we select the positive phase (1951, 1965, 1972, 1982, 1983, 1987, and 1997) with the composite index more than or equal to 3.8 and the quasi-normal phase (1952, 1956, 1960, 1967, 1968, 1979, 1980, 1981, and 1990) with the index near zero. For comparing with the pure ENSO mode, the pure El Niño year is chosen when the west–east difference of SSTA in the Indian Ocean (the Indian Ocean dipole index) is small: 1951, 1953, 1957, 1963, 1965, 1969, 1976, 1986, and 1991, and the quasi-normal year of the Pacific is also picked when the Niño$_{3.4}$ SSTA is about zero: 1959, 1960, 1962, 1980, 1981, and 1990.

The composite summer SSTA patterns of the above-mentioned types are shown in Fig. 2 respectively. The quasi-normal year of the composite index reflects the quasi-normal feature of SST better than that of the ENSO index. The SSTA in equatorial sea areas is very small for the quasi-normal year of the composite index. Moreover, there are a lot of differences between the positive phase year of the composite index and that of the ENSO index. The composite index displays the west–east difference of the tropical ocean SSTA, not only in the Pacific, but also in the Indian Ocean. These comparing results once again indicate that the tropic Pacific–Indian Ocean temperature anomaly composite mode considerably differs from the

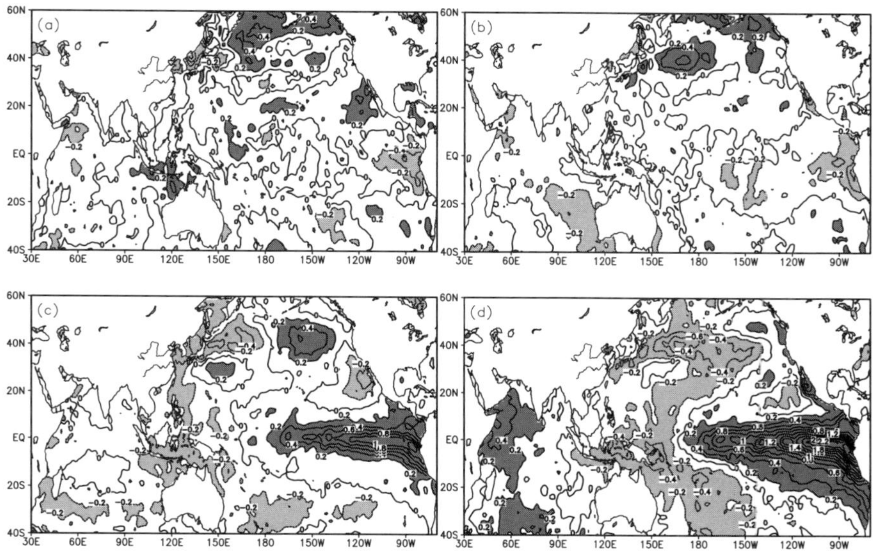

Fig. 2. Composite of SSTA in summer (JJA) for the quasi-normal phase of ENSO mode (a), the quasi-normal phase of the PIM (b), the positive phase (El Niño) of the ENSO mode (c), the positive phase of the PIM (d). Unit: °C. The anomalies greater than 0.2 are in dark shading and those less than −0.2 are in light shading.

Pacific ENSO mode, and proposing and studying the composite index is very significant.

3. Influence of the Composite Index on Summer Climate in Asia

3.1. *Rainfall and temperature in China*

Based on the corresponding Chinese climates for the four types of Fig. 2, we can easily find the influences of all SSTA patterns on climate are quite different. Because of long duration of SSTA, our results can provide some base for Chinese summer climate prediction. Because of the finite time length of observation data, the cases for composite analyzing are slightly small. However they, to a certain extent, explain the considerable effects of SSTA patterns on East China rainfall. Figure 3 shows the summer rainfall anomalies in East China corresponding to the four types of SSTA distribution in Fig. 2, respectively. Corresponding to the quasi-normal phase of the Pacific SSTA, more rainfall can be found in the coastal regions of

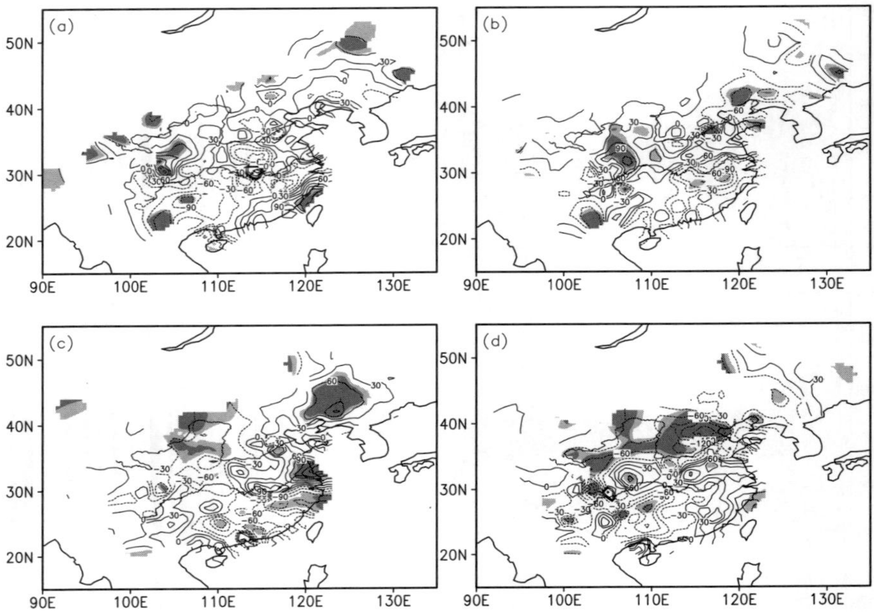

Fig. 3. Composite of summer (JJA) precipitation anomaly in China for the quasi-normal phase of ENSO mode (a), the quasi-normal phase of the PIM (b), the positive phase (El Niño) of the ENSO mode (c), the positive phase of the PIM (d). Unit: mm/JJA. The dark shading and light shading show the t-test significance at the 0.05 and 0.1 levels, respectively.

south and southeast China, the Changjiang–Huaihe River Basin, and the southern part of North China, but less rainfall is observed from south of the Changjiang River to Yunnan Province (Fig. 3(a)). Corresponding to the quasi-normal phase of the Pacific–Indian Ocean, south of the Changjiang River, North China and the southern part of Northeast China receive less rainfall, and the areas from the Changjiang–Huaihe River Basin to Sichuan Province record more rainfall (Fig. 3(b)). Corresponding to the pure El Niño year, there is more rainfall in the Changjiang–Huaihe River Basin and Northeast China, less rainfall to the south of Changjiang River (Fig. 3(c)). Corresponding to the positive phase of the PIM, we can find more rainfall in southeast China coastal region and from the Changjiang–Huaihe River Basin to Sichuan Province, and less rainfall in North China and Northeast China (Fig. 3(d)). Comparing Fig. 3(a) with Fig. 3(b), and Fig. 3(c) with Fig. 3(d), it clearly indicates that on the summer rainfall in

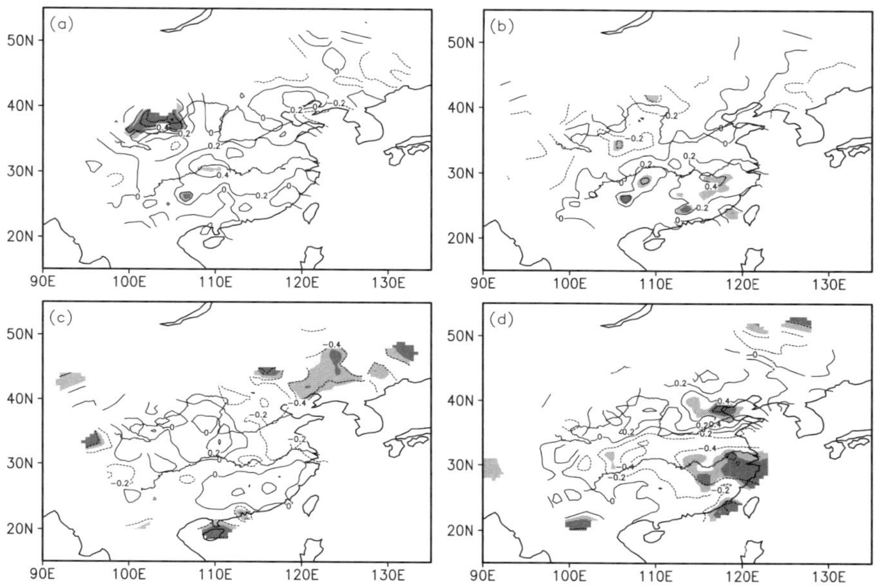

Fig. 4. Surface air temperature anomaly in China (unit: °C), with the rest being the same as Fig. 3.

China, the impacts of the pure Pacific SSTA are very different from the composite impacts of the Pacific–Indian Ocean SSTA.

In a similar way, Fig. 4 illustrates, respectively, the surface air temperature anomalies in East China for the four types of SSTA distributions of Fig. 2, featuring that corresponding to the quasi-normal year of the Pacific SSTA, the Changjiang–Huaihe River Basin gets noticeable positive temperature anomaly (Fig. 4(a)); corresponding to the quasi-normal year of the Pacific–Indian Ocean SSTA, South China receives remarkable positive temperature anomaly (Fig. 4(b)); corresponding to the pure El Niño year, Northeast China accepts substantially negative temperature anomaly; corresponding to the positive phase of the Pacific–Indian Ocean temperature anomaly mode, the middle-low region of the Changjiang River is controlled by negative temperature anomaly, and North China and the southern part of Northeast China have positive temperature anomaly (Fig. 4(d)). Evidently, the results indicate that the impacts of the pure Pacific SSTA on the summer temperature in East China are also different from those of the Pacific–Indian Ocean SSTA composite mode. In the summer of 2003, persistent severe hot weather has occurred in large areas of South China. The main reason is that the Pacific–Indian Ocean SSTA remains a quasi-normal state.[24]

Previous researches of the influence of El Niño on summer climates in China already showed that during El Niño year, more rainfall above average occurred in the Changjiang–Huaihe River Basin,[25,26] and low temperature and more rainfall in Northeast China.[27,28] Our results involving the pure El Niño are almost the same as those previously researched. Consequently, we can conclude that the analyses in this section are credible to a certain degree.

3.2. *Rainfall in India*

Although the relationship between the Indian summer precipitation and ENSO in recent years is some what weakened, on the whole, the summer anomalous rainfall in India is still a typical example of the ENSO impacts. One of the causes of the weakened relationship between the Indian summer rainfall and ENSO is considered the influence of the Indian Ocean SSTA. If the composite effect of the SSTA in both the Pacific and the Indian Ocean is taken into account, it may make the relationship between our presented composite mode and the Indian summer rainfall become relatively good. Figure 5 shows the Indian summer rainfall corresponding to the pure El Niño year and the positive phase of the PIM, respectively. More rainfall than average occurs in the middle part of India, less rainfall happens in the northern part and the southwestern part of India corresponding to the pure El Niño year (Fig. 5). The whole India receives substantially less rainfall than the average corresponding to the positive phase of the PIM (Fig. 5(b)).

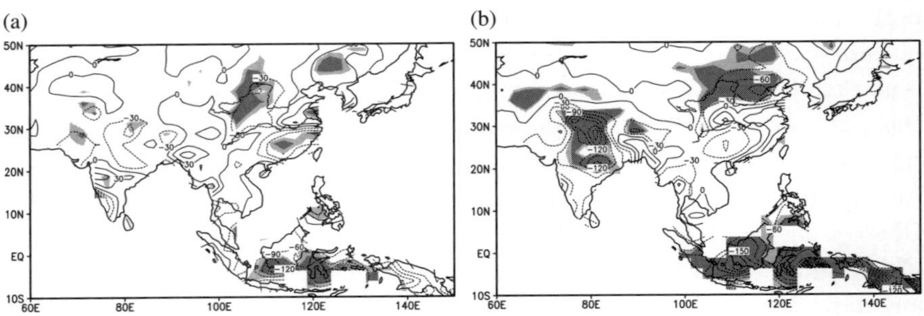

Fig. 5. Composites of precipitation anomalies from the precipitation reconstruction over land (PREC/L) in summer for positive phase of (El Niño) of the ENSO mode (a), (b) positive phase of the PIM. Unit: mm/JJA. The dark shading and light shading show the t-test significance at the 0.05 and 0.1 levels, respectively.

Comparing Fig. 5(a) with Fig. 5(b), it can be found that the difference between the influence of the pure El Niño and that of the positive phase of the PIM is quite large. The influence of the positive phase of the PIM can increase the drought in India. Although the mean summer rainfalls in 1983 (figure omitted) and 1997 are near average in India and slightly more than average in some regions of India,[29] other 5 years among the seven cases receive less rainfalls. It indicates that less rainfall in India is a main feature during the positive phase year of the PIM. Therefore, it is worth discussing that the occurring of the Indian Ocean temperature dipole causes the Indian rainfall in El Nino year being not negative anomaly. The relationship between ENSO and Indian rainfall is weakened recently much because of other factors including the role of Atlantic circulations.[30] In spite of the so-called "Weakening ENSO–Monsoon relationship" India faced severe droughts during the moderate El Nino episodes of 2002 and 2004. Studies have shown that the impact of ENSO events on Indian Monsoon rainfall is modulated by the decadal variability in monsoon rainfall and depends on the prevailing epoch; i.e. the impact of El Nino (La Nina) in more severe during the below (above) normal rainfall epochs.[32,33,35] The Indian Monsoon rainfall has been under the below normal epoch since last four decades, hence the impact of 2002 and 2004 El Nino episodes may have resulted in droughts over India.

The above observation analysis suggests that the PIM and its role are significantly different from the Pacific ENSO. For isolating the influence of SSTA on climate change, ensemble simulation is used for further investigating the climate effect of the PIM. The R42L9 numerical model in this paper is a global generation spectral model developed by State Key Laboratory of Numerical Modeling for Atmospheric Sciences and Geophysical Fluid Dynamics (LASG), Institute of Atmospheric Physics, Chinese Academy of Sciences. The horizontal resolution is roughly $2.8125(\text{lon}) * 1.66(\text{lat})$ with a rhomboidal truncation at wavenumber 42. The vertical adopts the terrain-following Sigma coordinate with nine levels. The model uses a reference atmosphere, and semi-implicit time-integration scheme with 15-min time step. This model was successful in modeling climate characteristics.[31]

Control experiments and sensitivity experiments have been performed with the time integration from 1 January to 30 September. The monthly mean SST is used in the control experiment. In the sensitivity experiment, the forcing SST uses the monthly mean SST plus the SSTA in the forcing region 15°S–15°N, 40°E–75°W which is twice that (Fig. 2(d)) of the positive

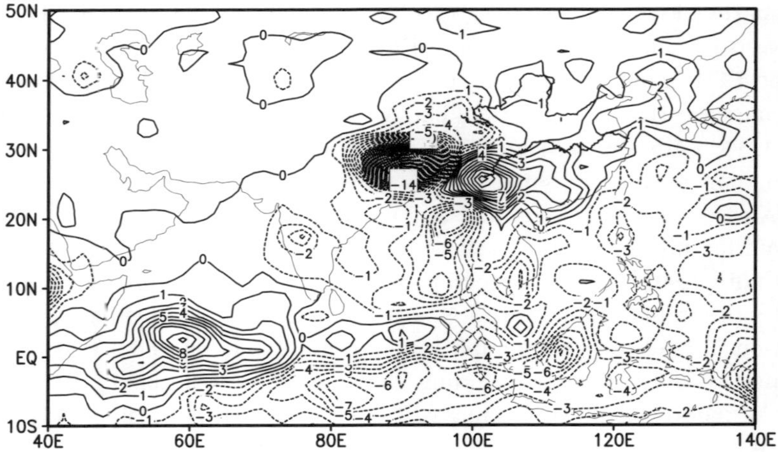

Fig. 6. Precipitation anomalies in summer (JJA) by the SSTA forcing of positive phase of the PIM. Unit: mm/d.

phase of the PIM after 16 April. In order to decrease influences of initial error and other errors, the control experiment and sensitivity experiment apply seven ensemble members starting from some different initial states. The seven ensemble members mean is used. From the difference between the control experiment and the sensitivity experiment, it can be the anomalous response to the SSTA forcing.

Figure 6 gives the precipitation anomaly forced by the SSTA of positive phase of the PIM. We can find that substantial negative anomaly occurs in Indian Peninsula, Indochina, and Indonesia, which is quite consistent with Fig. 5(b). But it differs from Fig. 5(a) of ENSO. The numerical results can be compared well with the observation evidence. The positive rainfall anomaly occurs from Southwest China to Northeast China with center in Sichuan Province and Guizhou Province of China and Korea in numerical simulation. The Changjiang–Huaihe River Basin and large part of South China also get more rainfall. This indicates that the numerical results in most area of China are the same as those observed. But in few areas of China the numerical results do not agree with those observed. The mechanism of summer precipitation in China is complex and middle-high latitudes are not all influenced by SSTA. Collectively, the numerical results are comparable well with those observed. The numerical simulation further indicates that the PIM has important impact on Asian climate which differs from that of ENSO mode.

4. Impacts of the Pacific–Indian Ocean Temperature Anomaly Mode on Asian Climate Systems

Figure 7 shows the composite of 100 hPa geopotential height for positive and negative phases (1954, 1955, 1964, 1970, 1988, and 1996 with the composite index less than or equal to 2.5) of PIM. The South Asian high is weaker and stretches eastward and southward for the positive phase. In contrast, the South Asian high is stronger and stretches westward and northward for the negative case.

Figures 8(a) and 8(b), respectively, show the composites for positive and negative phases of PIM at 500 hPa. In the case of positive phase, the contour of 5860 gpm of the western Pacific subtropical high is shifted to the coast areas of Fujian Province. The contour of 5880 gpm is found over the western Pacific to the east of Taiwan. During the year of negative phase, the contour of 5860 gpm retreats to the east of Taiwan Island. We cannot find the contour of 5880 gpm. So the PIM can influence the intensity and the longitudinal position of the western end of the western Pacific subtropical high.

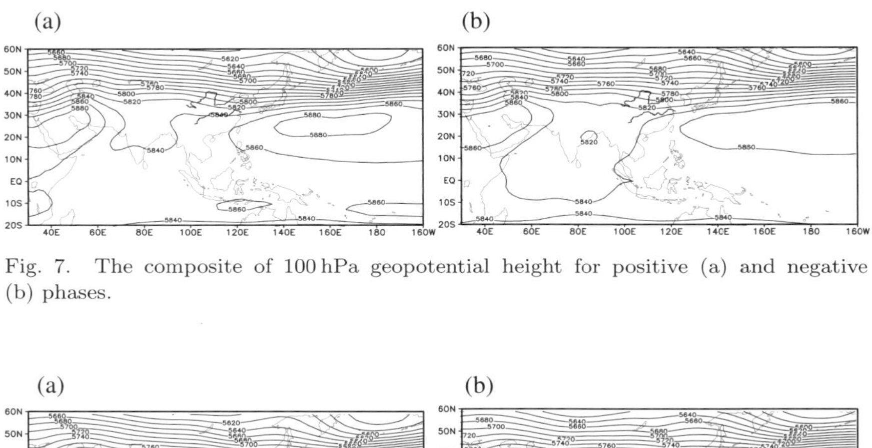

Fig. 7. The composite of 100 hPa geopotential height for positive (a) and negative (b) phases.

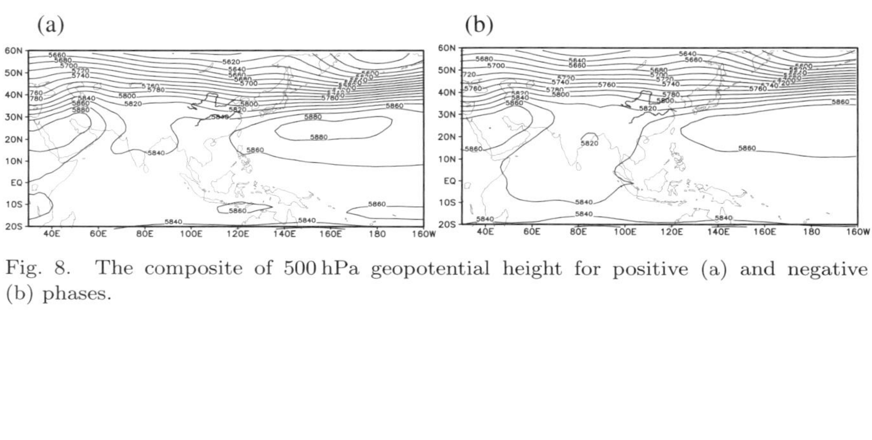

Fig. 8. The composite of 500 hPa geopotential height for positive (a) and negative (b) phases.

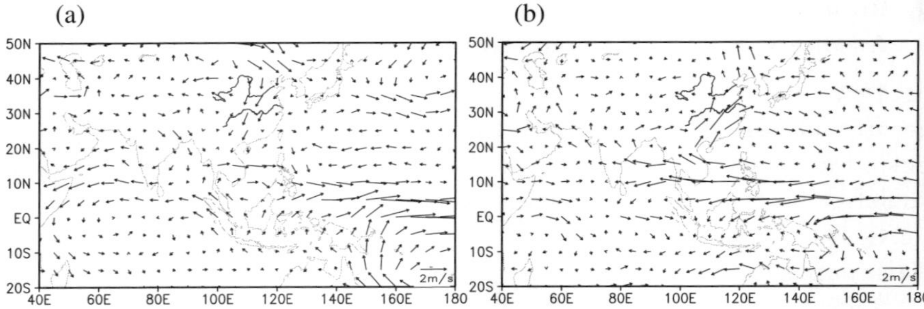

Fig. 9. The composite anomalies at 850 hPa for positive (a) and negative (b) phase.

The positive phase features the anomalous cross-equator southerlies from the eastern Indian Ocean to the western Pacific at 850 hPa (Fig. 9(a)). Thus, the South Sea monsoon trough and ITCZ are enhanced. Anomalous northerlies are found over East China. It indicates weaker East Asian summer monsoon. The Indian southwest monsoon is also weaker. In the year of negative phase (Fig. 9(b)), anomalous westerlies occupy over the western Indian Ocean, which means the stronger southwest monsoon. The anomalous southerlies over East China express the stronger East Asia summer monsoon. The anomalous cross-equator northerlies lie from the eastern Indian Ocean to the western Pacific. So the South Sea monsoon trough and ITCZ are weakened.

5. Conclusion

ENSO and Indian dipole should be regarded as an air–sea coupled system in the tropical Pacific and Indian Ocean. The Pacific–Indian Ocean temperature anomaly mode is presented. Comparing SSTA in the Pacific and Indian Ocean between the Pacific–Indian Ocean temperature anomaly mode and ENSO mode, we find that the quasi-normal year of the PIM can better reveal the quasi-normal feature of SST compared with that of ENSO mode. The SSTAs are all near zero around the equator. The differences in SSTA between the positive phases of the Pacific–Indian Ocean temperature anomaly mode and ENSO mode are also clear. The Pacific–Indian Ocean temperature anomaly mode can feature the east–west SSTA differences in both the Pacific and the Indian Ocean. It indicates that the Pacific–Indian Ocean temperature anomaly mode is different from the ENSO mode. The influences of the Pacific–Indian Ocean temperature anomaly mode on the precipitation and surface air temperature in China are very different from the ENSO mode. Thus, analyzing and forecasting the Asian climate change must consider the SSTA pattern in both the Pacific and the Indian Ocean, namely the tropical Pacific–Indian Ocean temperature anomaly mode. For the effect of SSTA on Indian rainfall, the PIM decreases much more Indian rainfall than pure El Niño. The explanation that the tropical Indian Ocean dipole is the cause of the weakening of the relation between ENSO and Indian precipitation in recent years may be not true. Other factors may be more important. The numerical simulation obtains generally the same results as the observational study. It proves that the PIM has important effect on Asian climate, which differs from ENSO mode. Therefore, to provide better scientific explanation for short-term climate prediction, the tropical Pacific–Indian Ocean temperature anomaly mode and its influence should be considered and investigated.

Furthermore, the characteristics of the Asian climate systems related to the Pacific–Indian Ocean temperature anomaly mode are investigated. The positive phase of the tropical Pacific–Indian Ocean temperature anomaly mode (positive SSTAs in the western Indian Ocean and eastern Pacific-negative SSTAs in the eastern Indian Ocean — western Pacific) is favorable for weaker Indian monsoon and eastern Asian summer monsoon, stronger South China Sea monsoon, a weaker South Asia high with southeastwards shift, a stronger subtropical high in the western Pacific with westwards move. The precipitation in southeast China coastal region and from the Changjiang–Huaihe River Basin to Sichuan Province increases, and the rainfall in North China and Northeast China decreases. The rainfall in India also decreases. The negative phase (reverse SSTAs of the positive phase) makes the southwest monsoon stronger over the western Indian Ocean and weaker over the India peninsula and the eastern Indian Ocean. It contributes to a stronger eastern Asian summer monsoon, weaker South China Sea monsoon, a stronger South Asia high with northwestwards shift, and a weaker subtropical high in the western Pacific with eastwards move. Less rainfall in Southeast China and more rainfall in the southwestern and northern China and to the south of Changjiang River is obtained. The precipitation in India increases.

Acknowledgments

This work was supported by the National Key Basic Research and Development Project of China (Grant no. 2004CB18300), and the Innovation Key Program of the Chinese Academy of Sciences (Grant no. ZKCX2-SW-226).

References

1. E. M. Rasmusson and T. H. Carpenter, The relationship between eastern equatorial Pacific sea surface temperatures and rainfall over Indian and Sri Lanka, *Mon. Wea. Rev.* **111** (1983) 517–528.
2. C. F. Ropelewsk and M. S. Halpert, Global and regional scale precipitation patterns associated with the El Niño/southern Oscillation, *Mon. Wea. Rev.* **115** (1987) 1606–1626.
3. R. H. Huang and Y. F. Wu, The influence of ENSO on the summer climate change in China and its mechanism, *Adv. Atmos. Sci.* **6** (1989) 21–32.
4. C. Y. Li, *Introduction to Climate Dynamics*, 2nd Edn. (Meteorology Press, Beijing, 2000) (in Chinese).
5. K. K. Kumar, B. Rajagopalan and M. A. Cane, On the weakening relationship between the Indian monsoon and ENSO, *Science* **284** (1999) 2156–2159.
6. C. Torrence and P. J. Webster, Interdecadal changes in ENSO–monsoon system, *J. Climate* **12** (1999) 2679–2690.
7. D. Z. Ye and R. H. Huang (eds.), *Study on the Law and Causes of Drought/ Flood in Yangtse River and Yellow River Valleys* (Shandong Scientific and Technological Press, Jinan, 1996) (in Chinese).
8. S. H. Luo, Z. H. Jin and L. T. Chen, Corrlation analyses of the SST in the middle-lower reaches of the Yangtze River, *Chinese J. Atmos. Sci.* **9**(3) (1985) 336–342 (in Chinese).
9. N. H. Saji, B. N. Goswami, P. N. Viayachandrom *et al.*, A dipole mode in the tropical Indian Ocean, *Nature* **401** (1999) 360–363.
10. P. T. Webster, A. M. Moore, J. P. Loschning *et al.*, Coupled ocean-atmospherie dynamics in the Indian Ocean during 1997–98, *Nature* **401** (1999) 356–360.
11. D. X. Wang, G. X. Wu and J. J. Xu, Interdecadal variability in the tropical Indian Ocean and its dynamic explanation, *Chi. Sci. Bull.* **44**(17) (1999) 1620–1626.
12. C. Y. Li and M. Q. Mu, Influence of the Indian Ocean dipole on atmospheric circulation and climate, *Adv. Atmos. Sci.* **18** (2001) 831–843.
13. Z. N. Xiao, H. M. Yan and C. Y. Li, The relationship between Indian Ocean SSTA dipole index and the precipitation and temperature over China, *J. Tropical. Meteor.* **18**(4) (2002) 335–344 (in Chinese).
14. J. H. He, R. H. Zhang, Y. K. Tan *et al.*, The features of the Interannual variation of sea surface temperature anomalies in the tropical Indian Ocean, in *Study on the Mechanism and Prediction of ENSO Cycle*, eds. J. P. Chao, C. Y. Li, Y. Y. Chen *et al.* (Meteorology Press, Beijing, 2003), pp. 279–293 (in Chinese).
15. C. Y. Li, M. Q. Mu and J. Pan, Indian Ocean temperature dipole and SSTA in the equatorial Pacific Ocean, *Chin. Sci. Bull.* **47** (2002) 236–239.
16. P. Niiler and J. Stevenson, The heat budget of tropical ocean warm-water pools, *J. Mar. Res.* **40**(Suppl) (1982) 465–480.
17. J. T. Potemra, R. Lukas and G. T. Mitchum, Large-scale estimation of transport from the Pacific to the Indian Ocean, *J. Geophy. Res.* **102** (1997) 27795–27812.

18. G. Meyers, Variation of Indonesian throughflow and the El Niño — Southern Oscillation, *J. Geophys. Res.* **101**(C5) (1996) 12255–12263.
19. Y. H. Pan and A. H. Oort, Correlation analyses between sea surface temperature anomalies in the eastern equatorial Pacific and the world ocean, *Climate Dyn.* **4** (1990) 191–205.
20. M. A. Alexander, I. Bladé, M. Newman *et al.*, The atmospheric bridge: The influence of ENSO teleconnections on air–sea interaction over the global oceans, *J. Climate.* **15**(16) (2002) 2205–2231.
21. X. S. Shen, M. Kimoto, A. Sumi, A. Numaguti *et al.*, Simulation of the 1998 East Asian Summer Monsoon by the CCSR/NIES AGCM, *J. Meteor. Soc. Japan* **79**(3) (2001) 741–757.
22. Y. F. Guo, Y. Zhao and J. Wang, Numerical simulation of the relationships between the 1998 Yangtze River valley floods and SST anomalies, *Adv. Atm. Sci.* **19**(3) (2002) 391–404.
23. Y. F. Guo, J. Wang and Y. Zhao, Numerical simulation of the 1999 Yangtze River valley heavy rainfall including sensitivity experiments with different anomalies, *Adv. Atm. Sci.* **19**(3) (2004) 391–404.
24. H. Yang and C. Y. Li, Diagnostic study of serious high temperature over South China in 2003 Summer, *Clim. Environ. Res.* **10**(1) (2005) 90–95 (in Chinese).
25. C. B. Fu and X. L. Teng, The relationship between the climate anomaly in China and El Niño/Southern Oscillation phenomenon, *Chinese J. Atmos. Sci.* (special issue) (1988) 133–141 (in Chinese).
26. D. Y. Gong and S. W. Wang, Impacts of ENSO on the seasonal Rainfall in China, *J. Natl. Disast.* **7** (1998) 44–52 (in Chinese).
27. Y. S. Liu, J. H. Zhi and Z. H. Zhou, The rule of period change of temperature and group occurring of low temperature in summer in Northeast China, in *Collected Papers of Long-Range Forecast of Summer Low Temperature in Northeast China*, Edit group of long-range forecast of summer low temperature in Northeast China (Meteorology Press, Beijing, 1983), pp. 17–21 (in Chinese).
28. C. Y. Li, El Niño event and the temperature anomalies in the eastern China, *J. Tropical Meteor.* **5** (1989) 210–219 (in Chinese).
29. G. Bell and M. Halpert, Climate assessment for 1997, *Bell. Am. Meteor. Soc.* **79** (1998) S1–S50.
30. C. P. Chang, P. Harr and J. Ju, Possible role of Atlantic circulations on the weakening Indian monsoon rainfall — ENSO relationship. *J. Climate* **14** (2001) 2376–2380.
31. T. W. Wu, P. Liu, Z. Z. Wang *et al.*, The performance of atmospheric component model R42L9 of GOALS/LASG, *Adv. Atm. Sci.* **20**(5) (2003) 726–742.
32. R. H. Kripalani and A. Kulkarni, Climatic impact of El Nino/La Nina on the Indian Monsoon: A new perspective, *Weather* **52** (1997) 39–46.
33. R. H. Kripalani and A. Kulkarni, Rainfall variability over Southeast Asia — Connections with Indian monsoon and ENSO extremes: New perspectives, *Int. J. Climatol.* **17** (1997) 1155–1168.

34. R. H. Kripalani, J. H. Oh, J. H. Kang, S. S. Sabade and A. Kulkarni, Extreme monsoons over East Asia: Possible role of Indian Ocean Zonal Mode, *Theoret. Appl. Climatol.* **82** (2005) 81–94.
35. K. E. Trenberth, P. D. Jones, P. Ambenje, R. Bojariu, D. Easterling, A. Klein Tank, D. Parker, F. Rahimzadeh, J. A. Renwick, M. Rusticucci, B. Soden and P. Zhai, Observations: Surface and atmospheric climate change, in *Climate Change 2007: The Physical Science Basis*, eds. S. Solomon, D. Qin, M. Manning, Z. Chen, M. Marquis, K. B. Averyt, M. Tignor and H. L. Millers, Contribution of Working Group I to the Fourth Assessment Report of the Intergovernmental Panel on Climate Change (Cambridge University Press, Cambridge, United Kingdom and New York, NY, USA, 2007).

Advances in Geosciences
Vol. 10: Atmospheric Science (2007)
Eds. J. H. Oh and G. P. Singh

BOUNDARY LAYER PHENOMENA OBSERVED BY CONTINUOUSLY OPERATED, TEMPORARY HIGH-RESOLUTION LIDAR

NOBUO TAKEUCHI, GERRY BAGTASA,
NOFEL LAGROSAS and HIROAKI KUZE
*CEReS, Chiba University, 1-33 Yayoi-cho, Inage-ku,
Chiba-shi, 263-8522, Japan*

SUEKAZU NAITO
*Chiba Prefecture Environmental Research Center,
1-8-8 Iwasakinishi, Ichikawa-shi, 290-0046, Japan*

MAKOTO WADA
*National Institute of Polar Research, 1-9-10, Kaga,
Itabashi-ku, Tokyo 173-8515, Japan*

AKIHIRO SONE and HIROFUMI KAN
*Hamamatsu Photonics, Inc., 5000 Hirakuchi, Hamakita-ku,
Hamamatu-shi, Shizuoka-ken, 434-8601, Japan*

TATSUO SHIINA
*Faculty of Engineering, Chiba University, 1-33 Yayoi-cho,
Inage-ku, Chiba-shi, 263-8522, Japan*

Continuous lidar observation of the atmosphere is important for monitoring various phenomena such as air pollution, local meteorology, and plume diffusion. Center for Environmental Remote Sensing (CEReS), Chiba University, developed a portable automatic lidar (PAL) with the cooperation of Hamamatsu Photonics Inc. After installing the automatic alignment capability, the PAL system has provided continuous observation data in every 20 s except for some maintenance periods. The lidar operates at 532 nm (second harmonic of Nd:YAG laser), with 1.4 kHz pulse-repetition frequency, 50 ns pulse width, and 15 μJ pulse energy. The signal is received by a 20-cm diameter telescope pointed northward with an elevation angle of 38°, and processed by a photon counter. The range resolution is 24 m (height resolution is 15 m). During the operation period of 3 years, we have observed atmospheric oscillations of Brunt-Vaisala type with oscillation periods of several to several 10 minutes, raindrops yielding information on falling speeds and corresponding droplet sizes, upward/downward motion of air in the boundary layer, and statistics on the cloud bottom height. In this chapter, we describe various features of

PAL data that characterize meteorological phenomena, including precipitation, frontal passage, and development of boundary layer.

1. Introduction

Since December 2002, we have operated a portable, automated lidar (PAL) that provides data on the vertical profile of the lower atmosphere every 20 s (Fig. 1). Continuous data with such a fine temporal resolution is valuable for systematic comparison between the lidar-observed features of the atmosphere and their meteorological interpretations. The micro-pulse lidar (MPL)[1] opened the way for continuously operating, automatic lidar systems. The network of MPL systems (MPLNET) is providing data on the boundary layer structure and cloud development at MPL sites. More recently, Sugimoto *et al.*[3] have constructed a lidar network consisting of 15 sites in east Asia, giving information on Asian dust with cooperation of other lidar sites in Japan (Asian dust NET, ADNET). Instrumentally, however, the MPL system often has difficulty in long-term operations because of the failure in the detector part resulting from its configuration (a single telescope is used for both transmitting and receiving). In an attempt to cope with this problem, we have adopted the design of the PAL in which the laser is installed on the side of the telescope, thus separating the transmitter optics and the receiving telescope.

Fig. 1. Photo of PAL.

The PAL system has been operated at Chiba Prefecture Environmental Research Center (CERC), located in an industrial park, nearby an air pollution monitoring station. The aerosol concentration data measured by a beta-ray instrument at the pollution-monitoring station is compared with the (optical) extinction data from PAL, yielding the value of mass extinction efficiency (MEE) of atmospheric aerosols in the boundary layer.[4,5] In this chapter, as a further extension of the PAL data analysis, we consider the relation between the PAL-observed features of aerosols/clouds and meteorological conditions that bring about such behavior of the atmosphere.

2. Monitoring System

The PAL is an eye-safe lidar with the laser energy of 20 μJ/pulse at 532 nm, pulse width of 50 ns, and repetition frequency of 1.4 kHz. It is installed on the second floor of the CERC (35.52N, 140.07E) building, and directed toward the north sky with an elevation angle of 38°. Detected in the photon counting mode, the lidar signal gives a vertical profile of aerosol/cloud concentration continuously every 20 s. The specification is given in Table 1. In order to suppress the background due to sky radiation, the receiving field-of-view (FOV) is limited to 0.2 mrad. Since this small FOV easily generates off-alignment of the system due to temperature change, the direction of laser beam is optimized every 15 min by adjusting the direction of the pointing prism.

Table 1. Specification of PAL.

Laser	Diode (LD)-pumped Q-switched Nd:YAG laser
Wavelength	532 nm
Laser pulse width	50 ns
Laser pulse energy	15 μJ/pulse
Pulse repetition rate	1.4 kHz
Telescope type	Schmidt–Cassegrainian
Telescope diameter	20 cm
Field of view	0.2 mrad
PMT detector	Hamamatsu:HPK-R1924P
Resolution	160 ns ($\Delta R = 24$ m)
Integration	20 s
Photon counting	SR430 (Stanford Research)
Remote control	Via Internet (ADSL modem)
Alignment interval	15 min
Pointing	Manual

The received photon number is counted and integrated by a scaler. The digitized signal is saved and displayed on the screen of a PC that also controls the total system. The PC, in turn, is remotely monitored and controlled through the LAN system from Chiba University.

3. Available Data

In this chapter, we use the following data: lidar (PAL) data, hourly meteorological data, and weather map of each day at 9:00 JST. Since December 2002, the PAL system has been operated continuously except for the maintenance time: March to May, 2004 and February to May, 2006. In this chapter, we focus on the data of 2004 and 2005.

Synoptic features of meteorological conditions can readily be seen from the weather map provided from the Japan Meteorological Agency (JMA). Hourly meteorological data of 143 observatories in Japan are displayed in JMA climate statistics: URL http://www.data.jma.go.jp/obd/stats/etrn/index.php (in Japanese). There, the following parameters are tabulated for every hour: temperature, precipitation, dew temperature, water vapor pressure, relative humidity, wind velocity, wind direction, and sunshine time duration. The descriptions of snowfall, snow amount, weather, cloud amount, and visibility at 9:00, 15:00, and 21:00 hrs, can also be found. We use the data from the weather station (Chiba weather station; 35.60N, 140.10E), nearest from the location of PAL (CERC). The weather map of Japan at 9:00 JST is available from "Every day weather map," http://www.data.jma.go.jp/fcd/yoho/hibiten/index.html. The locations of high and low pressures, fronts, and isobaric lines can be obtained from the map.

4. Data Processing of Lidar Data

PAL provides a vertical profile of aerosol every 20 s. The range-corrected signal, $X(R)$, is obtained as

$$X(R) = X^2 P(R), \tag{1}$$

where R is distance, $P(R)$ is the observed signal substrated by the background due to skylight and dark noise. The ratio of the extinction to the backscattering coefficient (lidar ratio) is assumed to be $S_1 = \alpha_1/\beta_1$ for aerosols, and $S_2 = \alpha_2/\beta_2 = 8.53$ (sr) for air molecules. Then, the extinction coefficient $\alpha_1(R)$ is obtained by the Fernald formula[6] as

$$\alpha_1(R) = -\frac{S_1}{S_2}\alpha_2(R) + \frac{X(R)\exp\left[2\left(\frac{S_1}{S_2}-1\right)\int_R^{R_f}\alpha_2(R')dR'\right]}{\frac{X(R_f)}{\alpha_1(R_f)+\frac{S_1}{S_2}\alpha_2(R_f)} + 2\int_R^{R_f}X(R')\exp\left[2\left(\frac{S_1}{S_2}-1\right)\int_{R'}^{R_f}\alpha_2(r)dr\right]dR'}. \tag{2}$$

In the present analysis, we assume $S_2 = 30\,\text{sr}$ for aerosols and $17\,\text{sr}$ for clouds. In Sec. 6, the PAL data are depicted using the time–height indication (THI): the abscissa shows time, the ordinate shows the height, and the signal intensity is shown in the brightness (color or gray) scale. Because of the elevation angle (38°), the height resolution is improved by a factor of $\sin 38° = 0.62$.

5. Features of Meteorological Data

Before comparing PAL lidar data and meteorological conditions, the aspects of average meteorological conditions in 2004 and 2005 are overviewed. Monthly average of weather parameters of Chiba observatory is shown in Figs. 2(a) and 2(b).

From Figs. 2(a) and 2(b), it is found that in 2004, the precipitation amount of October was exceptionally large, while in July it was quite small. Usually in Chiba, precipitation is small from November to May, and large from August to October (except September, 2005). From the graph,

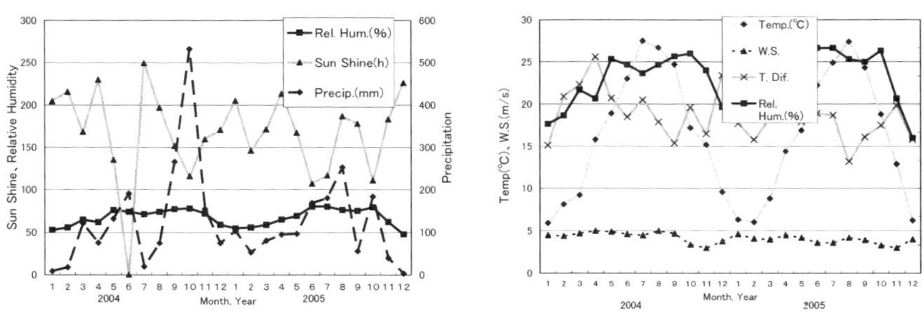

Fig. 2. Meteorological parameters at Chiba Observatory in 2004 and 2005. (a) Monthly precipitation (mm), sunshine hour (h), and relative humidity (%), (b) temperature (°C), relative humidity (%), wind speed (m/s), and temperature difference in a day (°C).

relative humidity exhibits some correlation (but not very significant) with precipitation and with temperature. Wind speed does not change so much annually, and the average speed is about 5 m/s. The average temperature in April and November is about 15°C: as mentioned later, the lidar data show different behaviors above and below 15°C, though the temperature difference (maximum temperature – minimum temperature) in a day stays between 15° and 25° throughout the year.

6. Classification Based on Meteorological Conditions

Although the diurnal behavior of relative humidity generally shows anti-correlation with diurnal temperature variation, the monthly average relative humidity exhibits positive correlation with the monthly average temperature (Fig. 2(b)). The inspection of lidar data indicates that the average temperature of 15° provides an important benchmark, in accordance with the classification of summer and winter in Japan. Therefore, in the following discussion, we treat summer and winter cases to highlight the characteristics. The following six cases will be treated below: (i) clear sky, (ii) rainy weather, (iii) sunshine and precipitation on the same day, (iv) cloud-top boundary layer, (v) frontal passage, and (vi) sudden change of cloud height. The last two cases correspond to variable weather conditions.

6.1. *Clear sky case*

On a clear day, PAL data gives information on distributions of aerosol particles. As the light-scattering of particles is strongly affected by relative humidity, lidar signal behavior strongly differs in summer and winter.

6.1.1. *Summer case*

Clear conditions continued from 14th to 16th in June, 2004. One-day PAL THI data on June 14, 2004 are shown in Fig. 3(a). From 6:00 JST, the boundary layer starts ascending and reaches 3 km at 15:00 JST, then, it descends down to 1 km in the evening. The visibility on this day was very good, ranging from 20 to 30 km. In the diurnal variation shown in Fig. 3(b), the difference in temperature is 9 K, but according to this variation, the relative humidity changes from 80% to 40%, with the small change of water vapor pressure from 19 to 13 hPa. The time of maximum temperature agrees well with the time of the highest mixing layer height.

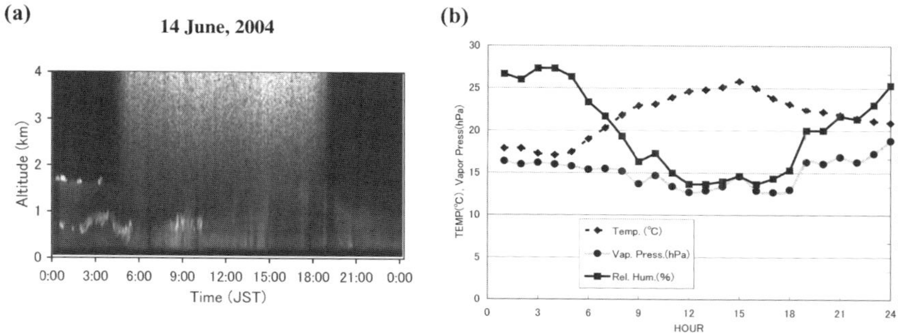

Fig. 3. Clear-sky case on 14 June, 2004. (a) Lidar data. From 6:00 to 15:00, the mixing layer develops and then it decays. (b) Variations of temperature, relative humidity, and water vapor pressure.

6.1.2. *Winter case*

In winter, both temperature and relative humidity are low. As an example of clear sky, the case of 3 February, 2005 is shown in Fig. 4. Lidar data in Fig. 4(a) show nearly constant aerosol concentration up to 4 km, and the development of mixing layer is not seen. The variations of temperature and relative humidity are shown in Fig. 4(b). The changes in temperature and relative humidity are from 0°C–10°C and from 55% to 20%, while the change in water vapor pressure is relatively small (from 3.5 to 2.3 hPa). The wind velocity becomes maximum in the daytime, and the direction is NNW with 10 m/s speed.

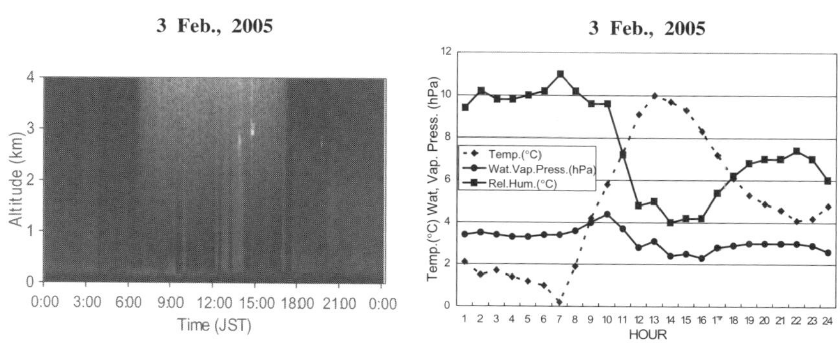

Fig. 4. Clear-sky case on 3 February, 2005. (a) Lidar data. No development of the mixing layer is observed. (b) The diurnal variation of temperature, relative humidity, and water vapor pressure.

6.2. *Rainy weather case*

During the precipitation, the cloud base almost touches the ground in both summer and winter. Nevertheless there is some seasonal difference in the observed lidar profiles.

6.2.1. *Summer case*

An example of 4 September, 2004 is shown in Fig. 5. The annually largest precipitation of 135.5 mm occurred on this day. Comparison of Figs. 5(a) and 5(b) indicates that the cloud indeed touches the ground during the precipitation period (2:00–3:00 and 19:00–24:00), when it is seen from the radar picture (not shown) that the cumulonimbus cloud covers Chiba prefecture moving from north-east to south-west. From 9:00 to 19:00, the cloud stays at a lower altitude (roughly 500 m), and the lidar signal that indicates the cloud base height fluctuates largely on a bad weather day. The weather map on this day shows that the stationary frontline stays off-shore of the Chiba prefecture.

6.2.2. *Winter case*

As an example of winter precipitation, Fig. 6 shows the data observed on 16 January, 2005. The total daily precipitation was 66.5 mm. The rain was almost over until 7:00. After that, the precipitation of roughly 1 to 2 mm lingered until the evening. The temperature was almost constant around 5°C–7°C.

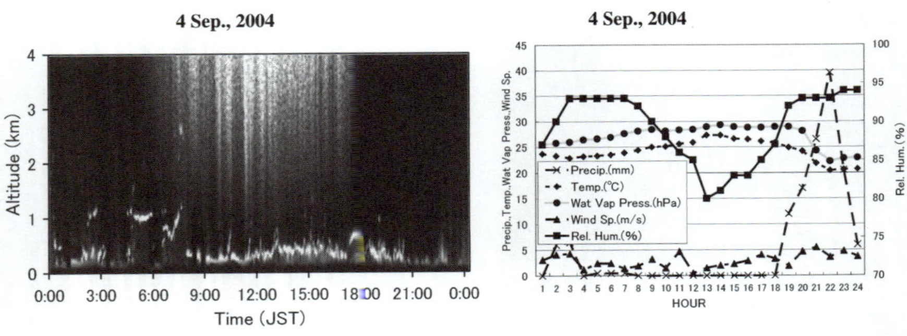

Fig. 5. Rainy weather case on 4 September, 2004. (a) Lidar data — The cloud base height touches the ground during the precipitation. (b) Diurnal variation of precipitation, temperature, relative humidity, water vapor pressure, and wind velocity.

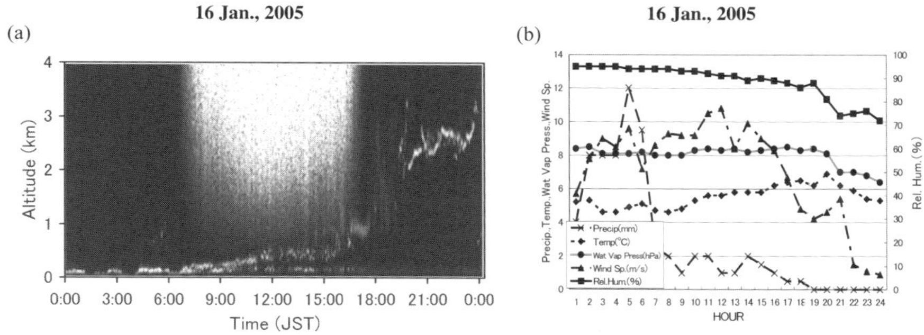

Fig. 6. Rainy weather on 16 January, 2005. (a) Lidar data — The cloud base height becomes lower during the precipitation. (b) Precipitation, temperature, relative humidity, water vapor pressure, and wind velocity.

6.3. *Sunshine and precipitation on the same day*

The features during sunshine and precipitation periods are largely different, due mainly to the difference in relative humidity and temperature. The transition in a relatively short time period leads to a large-scale, vertical fluctuation.

6.3.1. *Summer case*

Under such a situation, the cloud base height rapidly moves up and down. Figure 7 shows the lidar data on 7 September, 2004. Rapid fluctuation of cloud occurs in an altitude range between 0.5 and 2.5 km. On that day, a

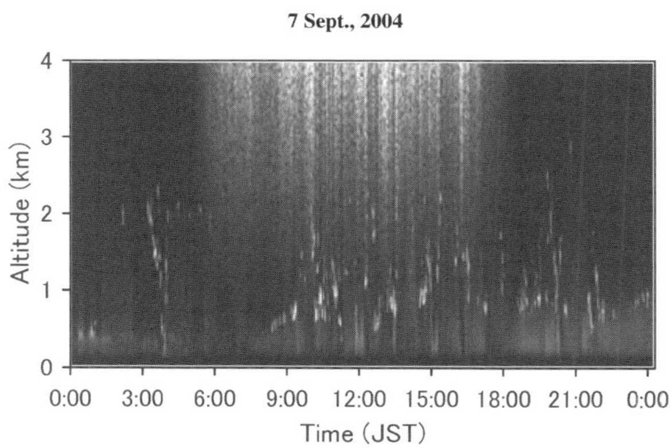

Fig. 7. Lidar data on 7 September, 2004.

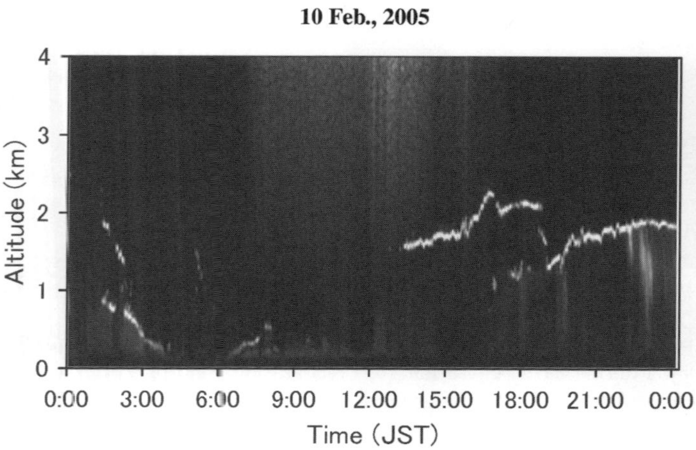

Fig. 8. Lidar data on 10 February, 2005.

typhoon was approaching, and the precipitation was 7.5 mm and the daylight time was 5 h.

6.3.2. *Winter case*

Figure 8 shows the PAL data on 10 February, 2005. The daily precipitation was 7.5 mm and daylight time was 4.3 h. The precipitation was recorded during 3:00–6:00 and 19:00–21:00. However, the relative humidity in the afternoon was less than 50%. This indicates that the precipitation evaporated before reaching the ground. Such a situation can often be seen in winter season. The rapid movement of the cloud base height is observed for both summer and winter cases when the sunshine and precipitation coexist on the same day.

6.4. *Appearance of cloud-top-boundary layer*

Boundary layers capped with cloud layers are frequently seen in the lidar data display. Such cases in summer (28 July, 2004) and in winter (25 February, 2005) are shown in Figs. 9 and 10, respectively. On 28 July, the precipitation and sunshine co-existed (hint of large fluctuation of cloud), while a typhoon was approaching. On 25 February, 2005, it rained until 17:00, and it became cloudy thereafter. The relative humidity was very high with a minimum value of 73%. Including these examples, all cases with

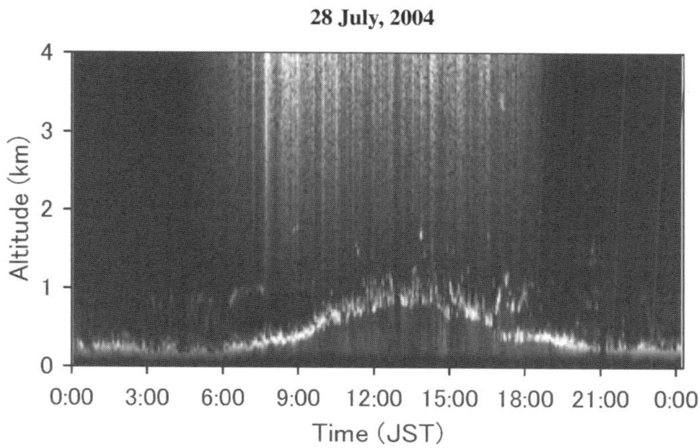

Fig. 9. Cloud-top-boundary layer on 28 July, 2004 (summer).

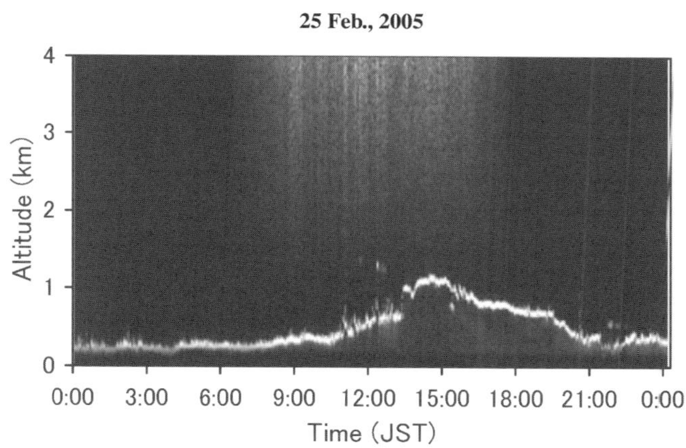

Fig. 10. Cloud-top-boundary layer on 25 February, 2005 (winter).

a cloud-top-boundary layer exhibit high relative humidity, with a diurnal minimum value of over 60%.

6.5. *Frontal passage*

In the case of frontal passage, one can expect the change of wind direction, gradual change of cloud height, and the starts of precipitation. However, various features are found in the lidar return signals, presumably due to the various types of frontal passage. Some typical cases are shown below.

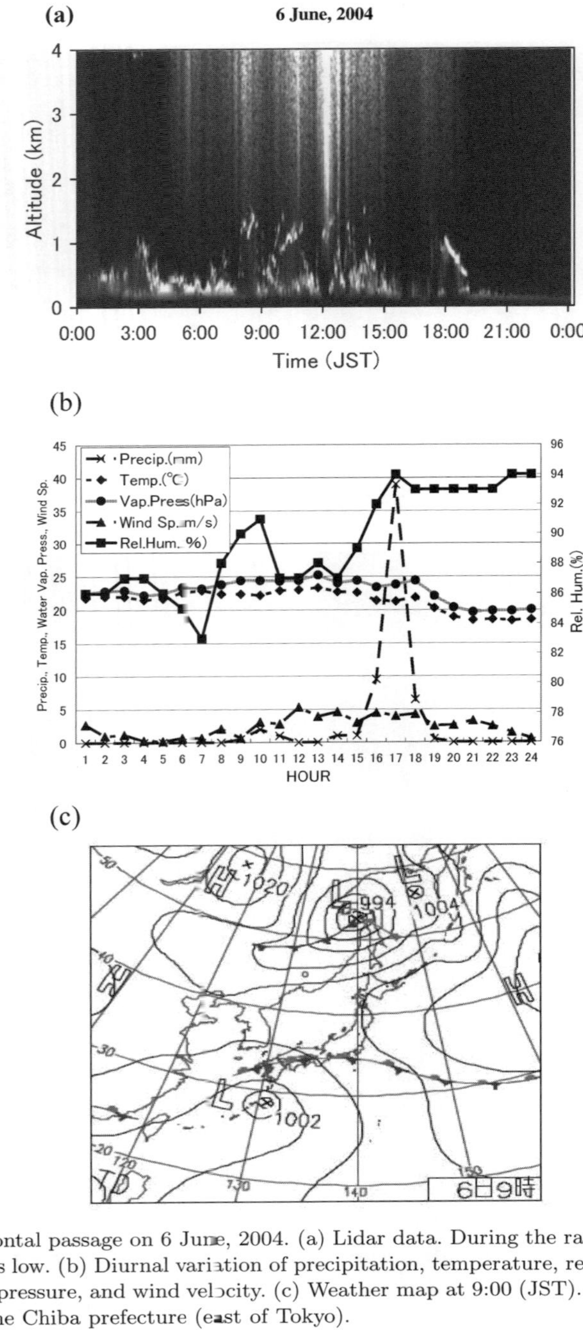

Fig. 11. Frontal passage on 6 June, 2004. (a) Lidar data. During the rainfall, the cloud base height is low. (b) Diurnal variation of precipitation, temperature, relative humidity, water vapor pressure, and wind velocity. (c) Weather map at 9:00 (JST). The front stays offshore of the Chiba prefecture (east of Tokyo).

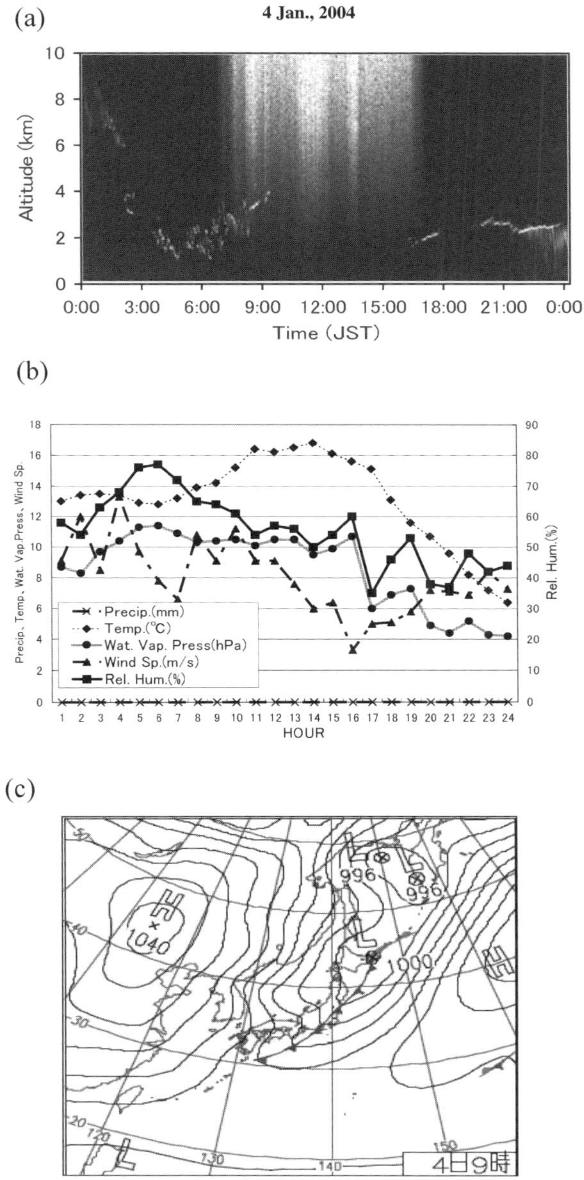

Fig. 12. Frontal passage on 4 January, 2005. (a) Lidar data. In the morning, cloud height sharply drops from 10 to 3 km. After 22:00, a cold front passes over the Chiba area. (b) Diurnal variation of precipitation, temperature, relative humidity, water vapor pressure, and wind speed. The maximum temperature appears around 15:00 due to the insolation in the daytime. (c) Weather map at 9:00 (JST). The front stays over the Chiba prefecture.

6.5.1. *A summer case*

A summer case on 6 June, 2004 is shown in Figs. 11(a)–11(c). On this day the precipitation was as much as 61 mm, and the onset of the rainy season was announced. From Fig. 11(b), a large amount of precipitation occurred during 16:00 and 18:00 with low cloud height. The rain stopped after 19:00. The wind direction changed from SE → E → NE → N. There was observed frequent jumps of the cloud bottom height.

6.5.2. *A winter case*

As an example in winter, a case in 4 January, 2005 is shown in Fig. 12. In Fig. 12(a), the vertical profile up to 10 km is shown. In the morning, cloud decreases the altitude suddenly from 10 to 1 km; then the cloud disappears and sunshine appears. Around 16:00, the wind direction changes from SW to NNE, and the temperature lowered sharply. Also, the relative humidity decreased from 80% to 30% due to the intrusion of cold, dry air. During 22:00 to 23:00, it is found that the raindrops evaporate even before reaching the ground.

6.6. *Sudden change of cloud height*

An example of sudden decrease of cloud height has already been mentioned in association with the frontal passage (Sec. 6.5). Similar phenomena are seen under different circumstances. Figure 13 shows a case in winter

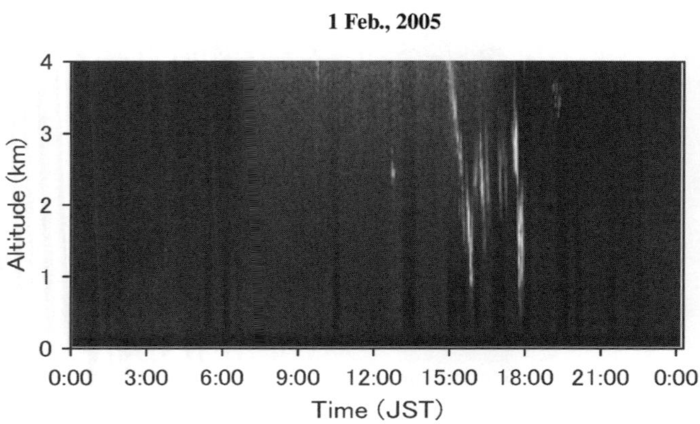

Fig. 13. An example of sudden change of cloud height observed on 1 February, 2005.

(1 February, 2005). There was no precipitation with the sunshine duration of 7.6 h, and wind direction changed from SW to NE with the speed of 3–12 m/s. The relative humidity was as low as 20% to 50%. The relation with the front is not clear.

7. Summary

In this chapter, we have analyzed the data of the automatically operated, temporary high resolution (20 s/file) lidar in conjunction with meteorological data obtained from a nearby weather observatory and public weather map. Important features can be summarized as follows: (1) Behavior of lidar data is strongly related to the relative humidity (RH). Daily average RH, in turn, exhibit noticeable correlation with the average temperature, although in a day, RH is negatively correlated with temperature. (2) Average ground temperature of 15° provides a good criterion separating the summer and winter modes of the atmospheric profile. The pattern of lidar data is classified into summer-type and winter-type, in accordance with the daily average temperature. (3) In winter-type data, RH is typically smaller. It is often observed that raindrops evaporate even before reaching the ground level. (4) Cloud-capped boundary layers are frequently observed. The condition for this phenomenon is the high RH, with a minimum RH value being over 60%. (5) When sunshine and precipitation coexist on the same day, the rapid, vertical movement of cloud base height is observed, presumably representing the instability of the atmosphere. (6) The following cases (for both summer and winter data) have been analyzed, and the lidar data have been compared with the meteorological conditions: clear day, precipitation, coexistence of sunshine and precipitation, cloud-capped boundary layer, frontal passage, and sudden change (drop) of cloud height. In future extension of this work, more quantitative treatment of the aerosol profile will give detailed understanding of the atmospheric processes that may lead to the observed features. The combination of lidar and concurrent radar data will improve the analysis of rain clouds.

References

1. J. D. Spinhirne, Micro pulse lidar, *IEEE Trans. Geosci. Remote Sensing* **31** (1993) 48–54.
2. http://mplnet.gsfc.nasa.gov/

3. http://www-lidar.nies.go.jp/
4. N. Lagrosas, Y. Yoshii, H. Kuze, N. Takeuchi, S. Naito, A. Sone and H. Kan, Observation of boundary layer aerosols using a continuously operated, portable lidar system, *Atmos. Environ.* **38** (2004) 3885–3892.
5. N. Lagrosas, H. Kuze, N. Takeuchi, S. Fukagawa, G. Bagtasa, Y. Yoshii, S. Naito and M. Yabuki, Correlation study between suspended particulate matter and portable automated lidar data, *J. Aerosol Sci.* **36** (2005) 439–454.
6. F. G. Fernald, Analysis of atmospheric lidar observations: Some comments, *Appl. Optics* **23** (1984) 652.

Advances in Geosciences
Vol. 10: Atmospheric Science (2007)
Eds. J. H. Oh and G. P. Singh

A MIE–RAYLEIGH-SODIUM FLUORESCENCE LIDAR SYSTEM FOR ATMOSPHERIC DETECTING

T. D. CHEN, X. H. XUE* and X. K. DOU
Department of Earth and Space Sciences, University of Science and Technology of China, Hefei, 230026, China

The Mie–Rayleigh-Sodium fluorescence lidar at the University of Science and Technology of China, Hefei (31.87°N, 117.23°E), China, has been set up in December 2005, designed for measuring atmosphere parameters at altitudes between ground level to 110 km. It is capable of detection of aerosol extinction (ground level to 30 km), atmospheric density, temperature (25 km–70 km), and sodium density (80 km–110 km).

During the past one year after the lidar was set up, we have carried on routine observations. The preliminary results of the sodium layer observation together with temperature profile and aerosol extinction profile are given in this chapter.

1. Introduction

The middle and upper atmosphere is a region of complex photochemical and dynamic interaction and is perhaps the least understood region of the earth's atmosphere. Laser remote sensing (lidar) is one of the powerful remote-sensing techniques which offer the greatest promise of high temporal and spatial resolution observations of the atmosphere. A large number of dynamic processes in the earth's atmosphere and long-period changes of atmosphere parameters can be explored using lidar.

In this chapter, we first give a description of the system configuration and technical characteristics of our multi-purpose lidar which measures vertical profiles of sodium density, atmospheric density and temperature, aerosol extinction over Hefei (31.87°N, 117.23°E). Then we present our preliminary observational results, i.e. the nocturnal and seasonal variations of the sodium layer, atmospheric density and temperature profile, and the aerosol extinction profile.

*Corresponding author.

Table 1. USTC lidar parameters.

	Nd:YAG	Dye laser
Transmitter		
Wavelength (nm)	532	589
Pulse energy (mJ)	550	50 (typ.)
Line width (cm^{-1})	1	0.05
Pulse width (ns)	6	6
Repetition frequency (Hz)	20	20
Divergence (mrad)	0.5	0.5
Receive-telescope		
Type	Cassegrain	
Aperture (mm)	1000	
FOV (mrad)	0.2–2	
Receiver-filter		
Wavelength (nm)	532	589.0
Bandwidth (nm)	1.0	0.5
Transmission	⩾60%	70%

2. Lidar System Configuration

A Mie–Rayleigh-Sodium fluorescence lidar has been assembled, tested and deployed to the University of Science and Technology of China (USTC) at Hefei, China in December 2005. The transmitter consists of a Nd-YAG laser at 532 nm and a Dye laser at 589 nm. The receiver is a 1 m diameter telescope. Parameters in detail are given in Table 1. The system has three channels which are used for different atmospheric parameters detecting, namely, sodium density, atmospheric density and temperature and aerosol extinction. Auto-align system is also employed to improve the system's automation. Two wavelength calibration systems can make sure the wavelength of dye laser output is near to the location of the maximum efficiency of sodium excitation.

3. Preliminary Experimental Results

3.1. *Na layer observation*

From December 2005 to April 2007, 53 night observations were obtained. Most of the observations cover the period of 20:00LT–04:30LT. The raw data consists of a series of photon counts corresponding to consecutive range bins. Each profile has an integration time of 4 min with a vertical resolution of 75 m.

3.1.1. *Nocturnal variation*

A typical photocount profile is plotted in Fig. 1(a). The profile was obtained at Hefei during the night of 23 December 2005. The resonant scattering from the Na layer between 80–105 km is clearly evident. The wavelike structure in the layer is caused by the wind perturbation associated with low-frequency internal gravity waves. The non-zero count level above 110 km is caused primarily by background noise from scattered moonlight and starlight.[1] Fig. 1(b) shows the sodium density variations with the local time and altitudes from 20:15LT 22 December 2005 to 04:00LT 23 December 2005. We can find during the midnight of 22 December, the sodium density increased to a very high value ($\sim$7000 cm^{-3}), which is called sporadic sodium layer and considered to be linked to the formation of sporadic E,[2] or the atmospheric dynamic procession (tide, gravity waves).

3.1.2. *Seasonal variation*

To investigate the seasonal variation, 53 nights' observational data were selected (exclude the period of sporadic sodium layer). It has been proved that the unperturbed sodium layer is a Gaussian profile:

$$\rho_{\mathrm{Na}} = \frac{A_{\mathrm{Na}}}{\sqrt{2\pi}\sigma_{\mathrm{Na}}} \exp\left[\frac{-(z - z_{\mathrm{Na}})^2}{2\sigma_{\mathrm{Na}}^2}\right]$$

where A_{Na} is column abundance, z_{Na} is centroid height and σ_{Na} is RMS width.

Seasonal variations of sodium column abundance (a), centroid height (b) and RMS width (c) are given in Fig. 2. Triangles stand for daily mean, error bars are daily standard deviations and dash lines are monthly mean values. Sodium column abundance reaches a maximum value of $6.014 \times 10^9\,\mathrm{cm}^{-2}$ in December and a minimum value of $1.126 \times 10^9\,\mathrm{cm}^{-2}$ in June. It reveals a significant summertime depletion followed by a maximum concentration in wintertime, especially in December. The monthly mean column abundance profile shows a minimum value in April due to that there is only one night data which lasted for 1 h in April, and we use it to fix the data gap during April. The same seasonal variations in abundance have been reported in other lidar sites,[3,4] and our results are consistent with those previous studies. Daily centroid height varies from 91.5 km to 92.5 km and has no obvious seasonal variation generally. It is higher in summertime and lower in wintertime. RMS width has a semiannual variation. It increases in summer and wintertime, decreases in spring and

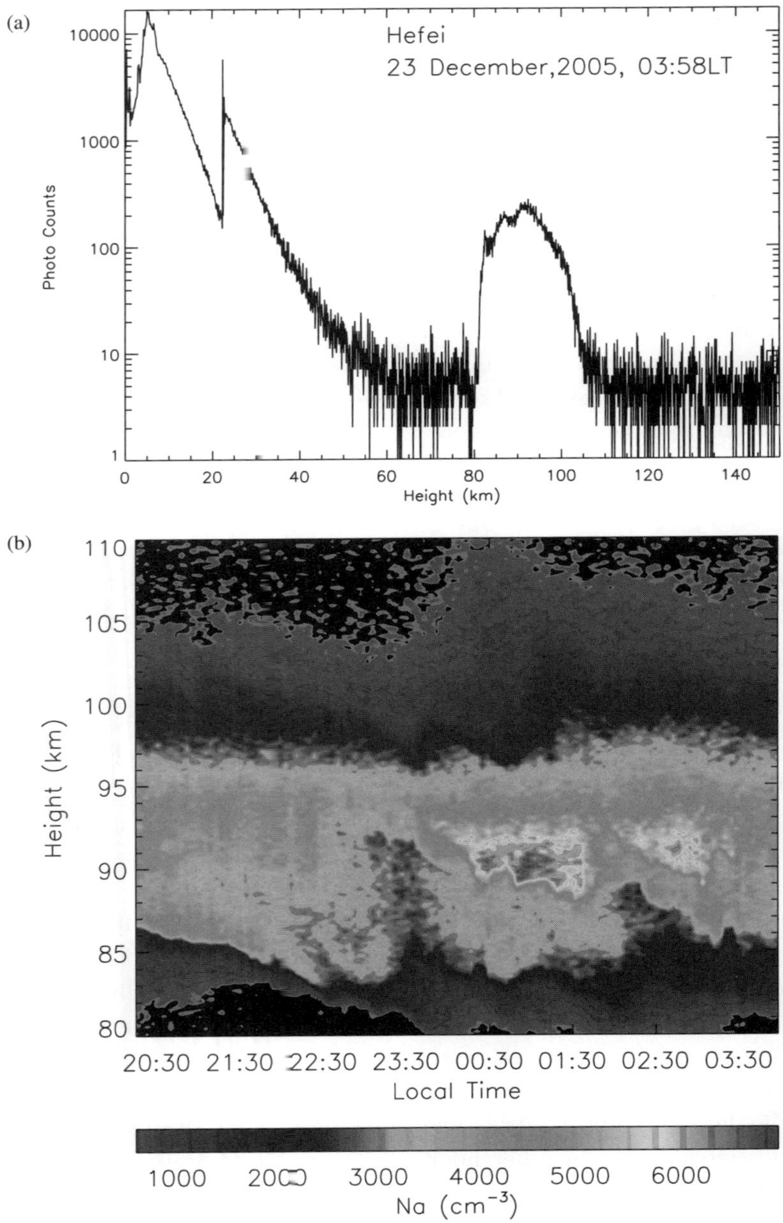

Fig. 1. (a) A typical lidar photocount profile at 589.0 nm, integrated time 4 min, with a vertical resolution of 75 m, at the night of 23 December 2005; (b) Sodium density variation during the night of 23 December 2005.

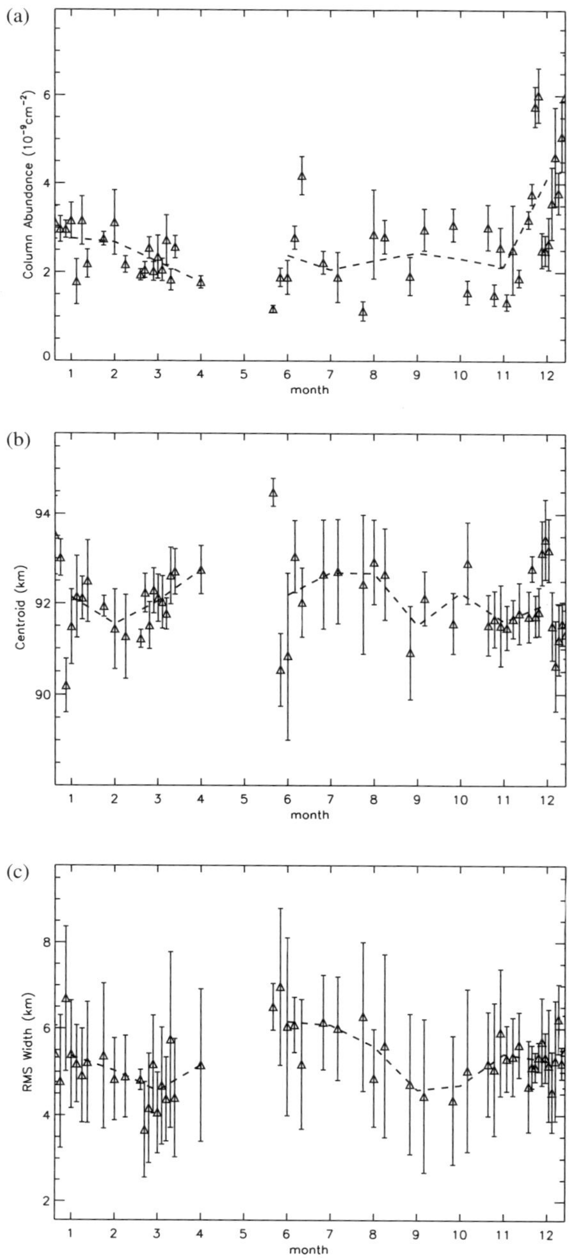

Fig. 2. Seasonal variations of sodium column abundance (a), centroid height (b) and RMS width (c). Triangle stands for daily mean, error bar is standard deviation, dash line is monthly mean.

autumn, i.e. it's 6.61 and 5.37 km in June and January, and 4.56 and 4.58 km in March and September, respectively.

3.2. *Atmosphere density and temperature observation*

In Rayleigh mode, the photocounts are integrated for 500 s with a vertical resolution of 150 m. The density and temperature profiles are inferred from the relative atmosphere density profile by using the hydrostatic equation and the ideal gas law. Figure 3 shows an example of the observation during 19:26–22:02LT, 6 December 2005. The solid lines represent the observational results and the dash lines show the monthly mean results of density and temperature obtained from CIRA-86. Below 70 km, the temperature difference between the observation and CIRA-86 is less than 10 K, while above 70 km is much bigger. This may due to the mesospheric temperature inversion.

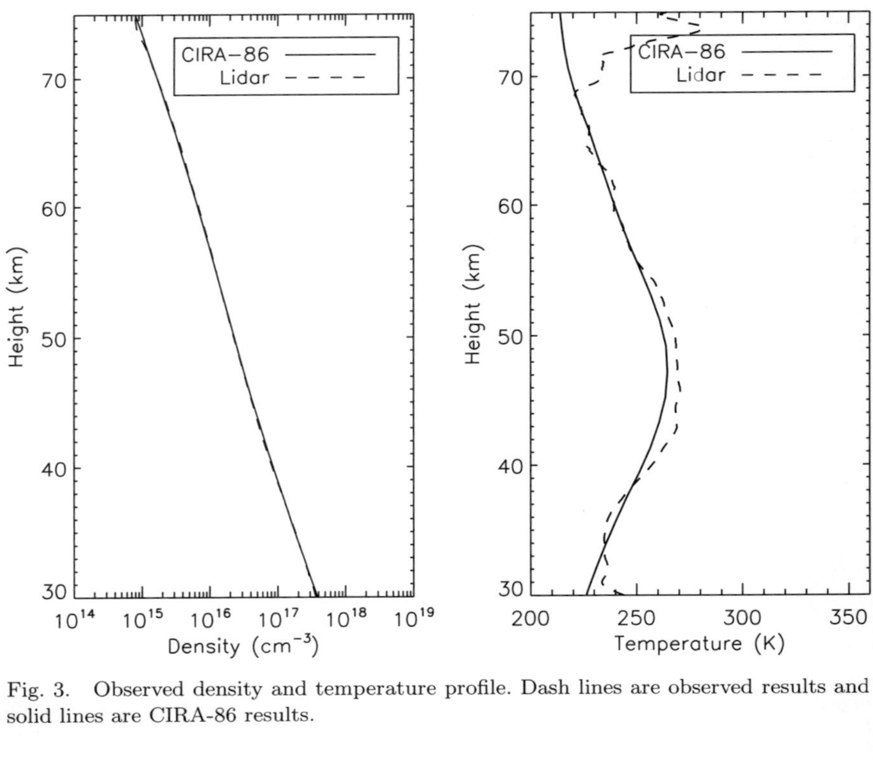

Fig. 3. Observed density and temperature profile. Dash lines are observed results and solid lines are CIRA-86 results.

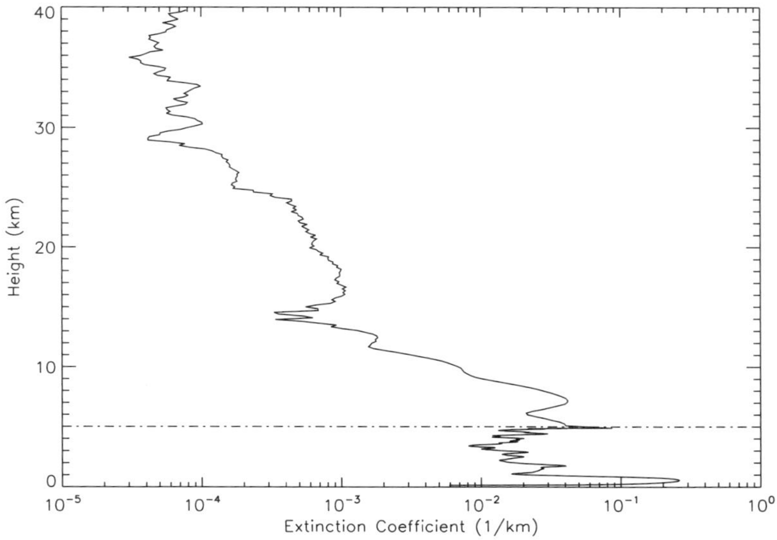

Fig. 4. Aerosol extinction profile obtained on the night of 1 June 2006.

3.3. *Aerosol observation*

Figure 4 shows co-observation result of stratosphere and troposphere aerosol extinction. The profile above the dash dot line (5 km) is obtained from stratospheric Mie-scatter channel with an integration time of 5 min and a vertical resolution of 150 m, the profile below the line is obtained from tropospheric Mie-scatter channel with an integration time of 5 min and a vertical resolution of 15 m. The detecting results of two channels agree with each other well at 5 km.

4. Conclusion

The Mie–Rayleigh-Sodium fluorescence lidar in USTC now has the capability to carry on Mie-scatter, Rayleigh-scatter and sodium resonance fluorescence measurements from ground level to upper atmosphere. Its technologic characters such as auto-align, auto-scan wavelength calibration have greatly improved the system's performance and automation.

Compared with the previous studies and the modeling results, it shows that our lidar system can give relatively reliable detection for sodium density, atmospheric density and temperature and aerosol extinction.

More information which reveals the atmospheric dynamics over Hefei will be studied further.

Acknowledgements

This work is supported by the KIP Pilot Projects of Chinese Academy of Sciences (KZCX2-YW-123, KGCX3-SYW-408), the National Natural Science Foundation of China (40674087, 40474052), the China Meteorological Administration Grant (GYHY200706013), the Cooperation Projects of IAP (IAP07307) and WHU (L06-3).

References

1. C. S. Gardner, Sodium resonance fluorescence Lidar applications in atmospheric science and astronomy, *Proc. of the IEEE* **77** (1989) 408–418.
2. P. P. Batista, B. R. Clemesha, I. S. Batista *et al.*, Characteristics of the sporadic sodium layers observed at 23 degree S, *J. Atmos. Sol. Terr. Phys.* **64** (2002) 15349–15358.
3. R. J. States and C. S. Gardner, Structure of the mesospheric Na layer at 40°N latitude: Seasonal and diurnal variations, *J. Geophys. Res.* **104** (1999) 11783–11798.
4. A. J. Gibson and M. C. W. Sandford, The seasonal variation of the night-time sodium layer, *J. Atmos. Terr. Phys.* **33** (1971) 1675–1684.

Advances in Geosciences
Vol. 10: Atmospheric Science (2007)
Eds. J. H. Oh and G. P. Singh

ANTHROPOGENIC AEROSOL RADIATIVE FORCING IN THE INDO-GANGETIC BASIN

SAGNIK DEY* and S. N. TRIPATHI†
Department of Civil Engineering, Indian Institute of Technology, Kanpur 208016, India
**sagnik@iitk.ac.in*
†snt@iitk.ac.in

Long-term (2001–2005) estimation of anthropogenic contribution to aerosol direct radiative forcing, and its spatial variability over Indo-Gangetic Basin (IGB) is presented. An optically equivalent model has been formulated based on the surface measurements of aerosol properties, and the optical properties are used to estimate the direct radiative forcing at the top-of-atmosphere (TOA), surface, and atmosphere. Anthropogenic aerosols contribute more than 80% to the composite aerosol optical depth (at 0.5 μm) in the winter, whereas the natural dusts contribute more than 55% in the summer. Mean annual clear-sky TOA, surface, and atmospheric forcing due to anthropogenic aerosols in the IGB are $+0 \pm 6.8$, -12.6 ± 6.7, and $+12.6 \pm 6.9\,\mathrm{W\,m^{-2}}$, respectively. Anthropogenic contribution is persistently found to be high in the eastern IGB (>70%). Anthropogenic aerosols contribute 55% to the mean (±SD) annual heating rate of $0.64 \pm 0.19\,\mathrm{K\,day^{-1}}$ over IGB; the heating rate even goes up to more than $0.9\,\mathrm{K\,day^{-1}}$ in many places seasonally. Persistently, large reduction of net surface radiation would affect the regional hydrological cycle through decrease in evaporation and sensible heat flux. The variable spatial heterogeneity in the atmospheric heating would affect the atmospheric circulation in this region.

1. Introduction

Aerosols perturb the Earth's radiation budget directly through scattering and absorption of sunlight,[1] and indirectly through interaction with clouds.[2] The change in the radiative fluxes at the top-of-the-atmosphere (TOA) and at the surface due to the aerosols (natural + anthropogenic) is termed as aerosol radiative forcing. One major uncertainty in aerosol direct radiative forcing (ADRF) arises from limited knowledge of the relative proportion of the anthropogenic and natural components and their mixing state.[3,4]

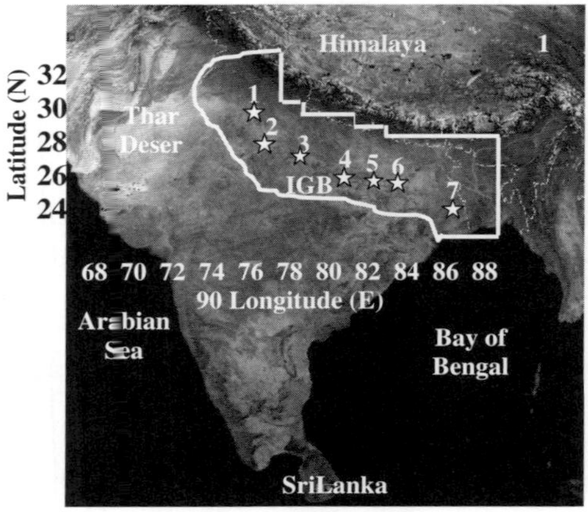

Fig. 1. The IGB in the northern India with the locations of major urban areas, as indicated by (1) Delhi, (2) Agra, (3) Kanpur, (4) Allahabad, (5) Varanasi, (6) Patna, and (7) Kolkata.

The Indo-Gangetic basin (IGB) in India (Fig. 1) is a region, where anthropogenic and natural aerosols have changing pattern in their loadings in different seasons.[5–7] In terms of aerosol loading, despite being one of the most polluted regions in the world,[8–12] the spatio-temporal distribution of ADRF in the IGB is poorly understood. Most of the previous studies are limited in spatial and temporal scale, e.g. Dey and Tripathi[12] and Tripathi *et al.*[9] have studied the aerosol and BC radiative effects in Kanpur (Fig. 1) during the winter months (December–February) based on the direct measurements of aerosol parameters during a land campaign conducted by Indian Space Research Organization in December 2004–January 2005.[11,13] Aerosol DRF estimation have been carried out for Delhi,[14] Hissar (semi-urban site in IGB),[15] and Nainital (high altitude site in the Himalaya),[16] but only for wintertime. Ramanathan and Ramana[8] have also investigated the climatic impact of absorbing haze over the IGB during the dry months (October–May). During the pre-monsoon (March–May) dust loading season, ADRF estimation has been carried out only for Kanpur[7] and Delhi[17] regions. All these studies indicated a very high magnitude of surface cooling (i.e. solar dimming) and atmospheric absorption (heating effect) over the respective locations, which has adverse effect on the regional climate. But how much of it is the anthropogenic fraction is not known, neither its spatio-temporal variability. Here, we report the spatio-temporal variations of the anthropogenic contribution to the ADRF over the IGB using ground-based and satellite observations, and through a regional aerosol optical model.

2. Methodology

The radiometers deployed at various locations around the world in a network, e.g. Aerosol Robotic Network, AERONET,[18] MWR network by Indian Space Research organization,[19] and the satellites measure the composite aerosol optical properties. To infer the anthropogenic contribution to the ADRF, one needs to delineate the anthropogenic components from the natural ones in the composite aerosol properties. An optically equivalent aerosol model constrained by ground-based direct and retrieved measurements of various aerosol parameters has been developed to achieve this. In this section, first the model features are described, followed by the method of estimating anthropogenic fraction and the uncertainties involved in these estimations.

2.1. *Model features*

The model considers three major individual components, water-soluble (gas-to-particle conversion and organics), black carbon (BC), and mineral dust, based on the chemical composition analyzed in Kanpur during the winter (December–February) season.[13] The main idea is to have a unique combination of these components for any particular time, which can reproduce the composite optical properties similar to those retrieved by AERONET. In this way, information on the individual components can be obtained. To achieve this, first we calculated the BC number concentrations from the measured BC mass concentration during December 2004 onwards.[9] For the earlier period (January 2001–November 2004), we have considered the BC concentration inferred from the AERONET-retrieved refractive index (for the algorithm, see Ref. 20). Once the BC number concentration is fixed, the number concentrations of the other components are varied iteratively until the model-derived composite aerosol optical properties match with the AERONET-retrieved optical properties. During the simulations, the aerosol vertical profiles are taken from the aircraft measurements by Tripathi *et al.*[10,21] over Kanpur during the winter, pre-monsoon, and post-monsoon (October–November) seasons. Another factor considered here is the relative humidity (RH). The model simulates the optical properties as a function of RH and we have compared the model results at RH closest to the ambient RH measured by an automatic weather station in Kanpur for that particular month. Comparisons of model-derived spectral AOD, single scattering albedo (SSA), and asymmetry parameter (g), and the Angstrom exponent within ±5% (which is the AERONET's retrieval uncertainty) of the AERONET-retrievals are considered for desired match.

2.2. *Estimation of the anthropogenic fraction in ADRF*

The mean annual SSA (at $0.5\,\mu$m) for anthropogenic aerosols is 0.89 with the lowest value (0.86) in December. During this time (i.e. winter season), the anthropogenic aerosols contribute more than 80% to the AOD and more than 90% to the total aerosol mass. The relative contributions of the anthropogenic aerosols to the composite aerosol properties decrease from March onwards due to commencement of dusts in the basin. The optical properties of the anthropogenic aerosols and composite aerosols derived from the optical model are used as input to the SBDART[22] to estimate the ADRF and the anthropogenic ADRF over Kanpur region first.[23] Then from the estimates of composite (anthropogenic + natural) aerosol forcing (ΔF) and anthropogenic ADRF (ΔF_{A}), we have followed the relation[3]:

$$\Delta F_{\mathrm{A}} = \Delta F * \mathrm{AEF} * \mathrm{AF}, \tag{1}$$

where AEF is anthropogenic efficiency factor and AF is the anthropogenic fraction. AEF is defined as the ratio of anthropogenic aerosol forcing efficiency to composite aerosol forcing efficiency, forcing efficiency being known as the DRF per unit optical depth. Both the surface and atmospheric ADRF are used to obtain two sets of AEF values for each month.[23]

Next, the aerosol forcing efficiency calculated from the estimations is assumed to be the representative of IGB,[12] and by multiplying the forcing efficiency with AOD from Moderate Resolution Imaging Spectroradiometer (MODIS) in $1^\circ \times 1^\circ$ grid, ADRF over the IGB in spatial scale is obtained. Over the IGB, the nature of aerosol loading is represented as monthly frequency distribution in Fig. 2(a) with the monthly mean (± standard deviation, SD) $\mathrm{AOD}_{0.55}$ value written in corresponding histogram. The distribution is unimodal in November to May and September, bi-modal in June, August and October, and tri-modal in July. The spread of the unimodal distribution in the winter season is narrower than the other

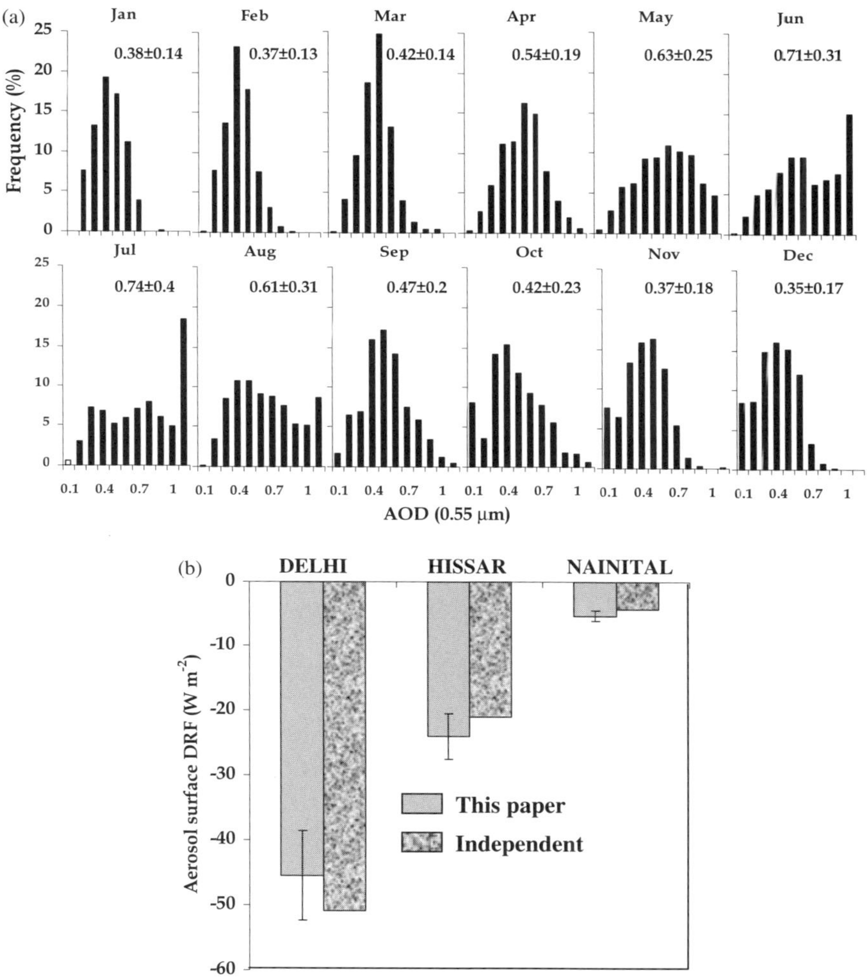

Fig. 2. (a) Frequency distribution of $AOD_{0.55}$ as derived from MODIS over IGB. Monthly mean (± standard deviation) AOD is shown at the top of each panel. (b) Comparisons of ADRF over few locations in IGB estimated using our model with the independent estimates.

months because during that period, aerosols are mostly anthropogenic and the nature of loading is uniform throughout the basin. In the summer months, dust activities start increasing and add to the anthropogenic burden, resulting in widening of the histograms and shifting of the peaks to the larger values. The spatial distribution of monsoon rainfall in the IGB is not uniform; at some parts it washes out the aerosols from the atmosphere through wet deposition, while in other parts aerosol loading remains high. This results in wide bi- and tri-modal distributions. In the post-monsoon season, the AOD distribution tends to readjust to unimodality and the frequencies of large AOD start decreasing. The ADRF estimated in the clear-sky condition using our model are compared with the independent estimates over some locations in the IGB to test the validity of our model (Fig. 2(b)), which shows that the expansion of our model is able to produce the ADRF within the current uncertainty levels. The cloudy-sky ADRF is calculated incorporating the cloud parameters from MODIS (for details see Ref. 23)

The critical part of Eq. (1) is AF, as no such measurements exist in IGB. We employ MODIS–aerosol fine mode fraction (AFMF) product to infer the AF. Earlier, researchers have utilized MODIS–AFMF data to assess the anthropogenic contribution over the oceanic region,[3] as in general, anthropogenic aerosols dominate in the fine mode fraction. The model-derived AF was compared with the MODIS–AFMF product over Kanpur for the entire 5 year period and a statistically significant relationship (correlation of 0.96) was found at 95% confidence level:

$$\mathrm{AF} = \mathrm{AFMF} * 0.65 + 0.1534. \tag{2}$$

The equation suggests that MODIS has a bias in determining the AF for this region, and thus using the above relation, AF was estimated for each $1^\circ \times 1^\circ$ grid of IGB. Subsequently, the anthropogenic ADRF at the surface and the atmosphere were estimated for each grid using Eq. (1). The anthropogenic ADRF at the TOA was calculated from the anthropogenic surface and atmospheric ADRF values.

2.3. *Uncertainty in anthropogenic ADRF*

The estimations of anthropogenic ADRF over the IGB involve some assumptions leading to uncertainty in our estimates, which we have quantified so that our results can be compared with other studies. There are two basic assumptions involved here, first all the water-soluble components are considered to be of anthropogenic origin, and second, the aerosol forcing efficiency over Kanpur is considered to be representative over the entire IGB. Uncertainties in the measurements and AERONET-retrievals, which are used to formulate and constrain the model, also propagate to increase the overall uncertainty. Dey and Tripathi[23] have presented a detailed analysis on the error budget of the estimations, where they have found that uncertainties vary in the range 5–20% due to various individual parameters. The overall uncertainty in the anthropogenic ADRF comes out to be ~21%.

3. Results and Discussion

3.1. *Spatio-temporal variations of ADRF in IGB*

Looking spatially for clear-sky ADRF, the central portion of IGB from west to east has highest surface ADRF in the winter ($<-30\,\mathrm{W\,m^{-2}}$) along the major urban areas. The values reduce in February in general with higher ADRF still visible in isolated patches in the western and easternmost regions. Surface ADRF starts building up again from March, reaching the maxima ($>-45\,\mathrm{W\,m^{-2}}$) in May–June. As monsoon arrives in the IGB, aerosols are being washed out and the reduced burden of aerosols decreases the surface ADRF until September. From October, aerosol loading again starts increasing and continues to winter. Mean ($\pm$SD) annual clear-sky TOA, surface, and atmospheric ADRF in the IGB are estimated to be -2.9 ± 4.3, -25.5 ± 13, and $+22.6 \pm 12\,\mathrm{W\,m^{-2}}$, respectively. In the presence of clouds, mean annual TOA ADRF switches from cooling to heating, while the surface ADRF becomes less negative. This has compensated the excess heating at TOA maintaining the atmospheric heating almost similar to the clear-sky condition. The reduction in magnitude of the negative surface ADRF in cloudy-sky condition is high (more than $10\,\mathrm{W\,m^{-2}}$) during the December–January and May–July. In these months, TOA forcing also increases by more than $7\,\mathrm{W\,m^{-2}}$. During the winter, as most of the aerosols are confined within the boundary layer, they get lesser chance to interact with the incoming solar radiation, a major part of which is reflected back to the space by optically thick clouds. Hence the surface cooling due to aerosols reduces and TOA ADRF flips toward positive side indicating warming effect. In August, the absolute magnitude of surface forcing is low due to least aerosol loading; hence the relative change in surface ADRF due to inclusion of cloud is also less. In other months, as the cloud fraction is low (<0.3), the effect is not so conspicuous.

The spatial heterogeneity in surface ADRF for four distinct seasons is shown in Fig. 3, where the latitudinally averaged ADRF values are presented with their standard deviations (plotted as vertical bars). Surface ADRF in winter and post-monsoon seasons display similar range of

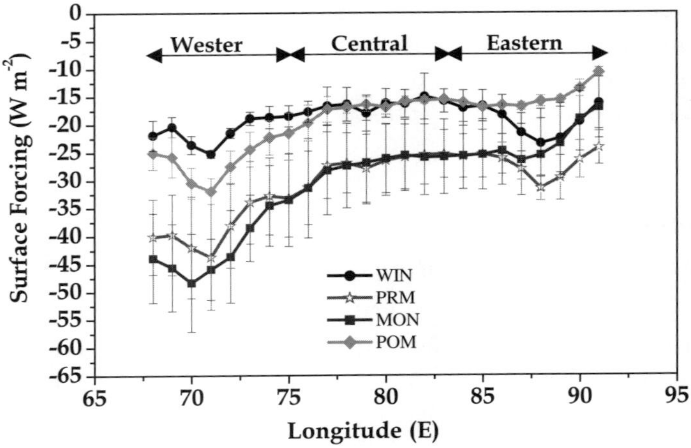

Fig. 3. Longitudinal variation of surface ADRF (presented as latitudinal average) in the IGB during the winter (WIN), pre-monsoon (PRM), monsoon (MON), and post-monsoon (POM) seasons. The vertical bars through each point represent the standard deviation for 2001–2005.

values and are distinctively lower than the corresponding pre-monsoon and monsoon values up to 85E longitude. Surface forcing further increases toward east of this longitude in three seasons except the post-monsoon season. The spatial variation is less in the winter and post-monsoon seasons, because these periods are dominated by anthropogenic aerosols, which is present throughout the basin. On the contrary, in the pre-monsoon and monsoon seasons, dusts are being transported over the IGB from the western arid regions influencing the optical properties.[5] The highest surface forcing in the western part of IGB is due to the strongest dust activities, which decreases eastward, as the dust loading reduces. Higher standard deviation in the dust loading seasons indicates higher heterogeneity across the latitudes, which depends on the transportation path of the dusts.

3.2. *Anthropogenic contribution to aerosol DRF over the IGB*

In this section, the variability of anthropogenic contribution to ADRF over IGB has been discussed with the term "anthropogenic forcing" used for "anthropogenic ADRF" for the sake of brevity. The relative contribution of the anthropogenic contribution to ADRF over Kanpur shows strong seasonal variation (Fig. 4). The mean annual relative contributions to

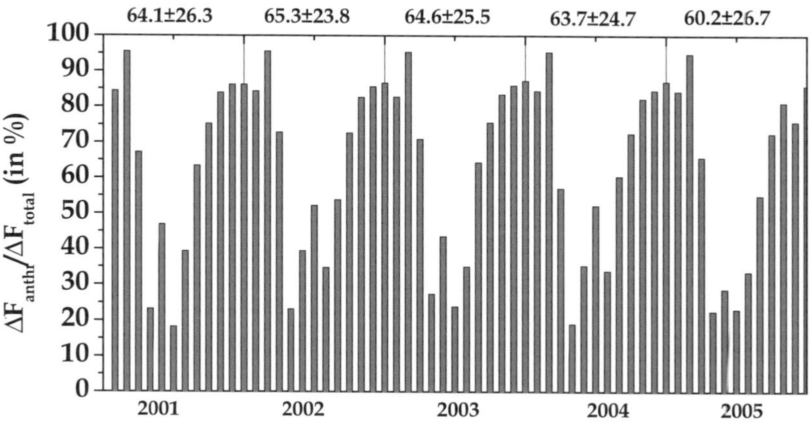

Fig. 4. Monthly contribution of anthropogenic aerosols to the total ADRF at surface over Kanpur (in percentage) for 5 years. The annual mean (±SD) value is written over each box.

surface ADRF are 64.1%, 65.3%, 64.6%, 63.7%, and 60.2% for 2001, 2002, 2003, 2004, and 2005, respectively. The large SD indicates the high variability within a year. In 2005, the proportion of anthropogenic aerosols to total aerosol surface forcing is lowest on annual scale. Further analysis based on seasons reveals that the proportion in the pre-monsoon season is lowest (37%) in 2004. In the years 2003 and 2005, the proportions are very low (<42%) in the monsoon season also. In the winter season, the anthropogenic contribution is uniform in all 5 years (88–89%), suggesting insignificant inter-annual variation.

The monthly mean spatial distributions of anthropogenic surface and atmospheric forcing over the IGB are illustrated in Fig. 5(a). The mean annual clear-sky TOA, surface, and atmospheric anthropogenic forcing in IGB are $+0 \pm 6.8$, -12.6 ± 6.7, and $+12.6 \pm 6.9\,\mathrm{W\,m^{-2}}$, respectively. The major difference in the spatial distributions of the anthropogenic surface forcing with total surface forcing is the persistent high values of anthropogenic forcing in the eastern IGB throughout the year except July–August, when monsoonal rain removes major fraction of aerosols from the atmosphere. However, the anthropogenic surface forcing over the urban centers like Kanpur and Delhi are still high ($>-15\,\mathrm{W\,m^{-2}}$), which shows that even in the monsoon season, the washout of aerosols by monsoonal rain is not so effective as compared to the eastern IGB. Immediately after the monsoon season, anthropogenic surface forcing starts building

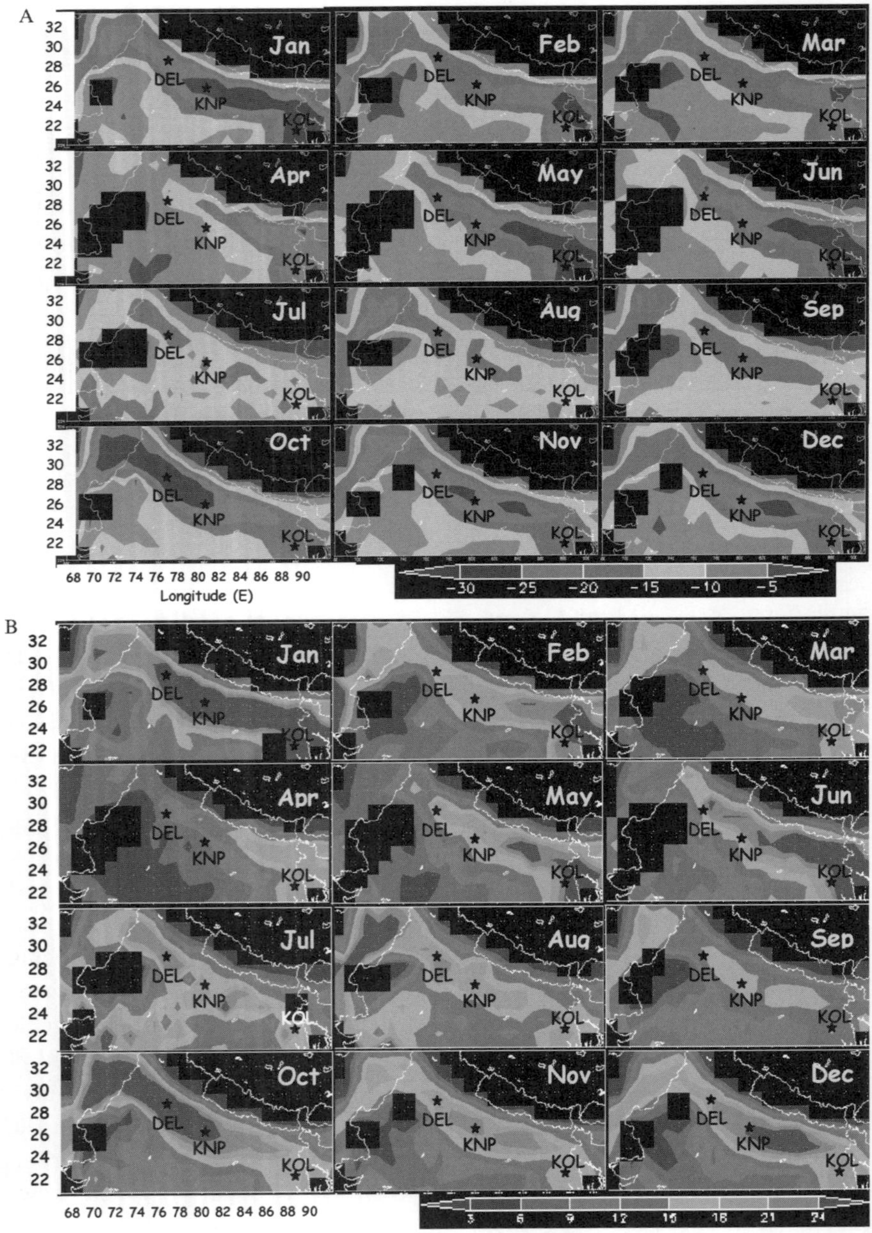

Fig. 5a. Monthly mean estimates of clear-sky anthropogenic aerosol (A) surface (top 12 panels) and (B) atmospheric (bottom 12 panels) forcing over IGB. Note that the color scales for the two sets of figures given are different.

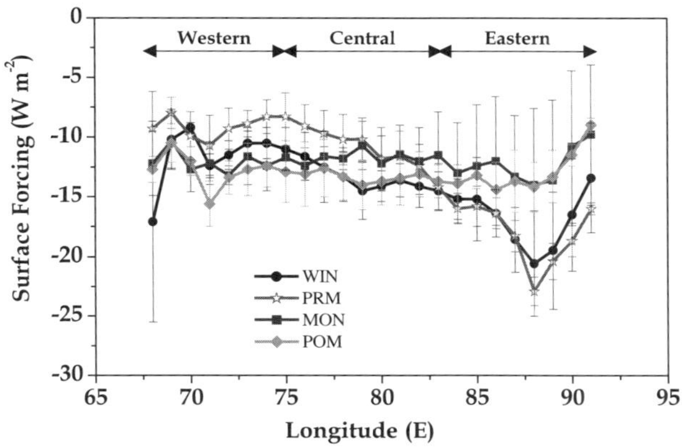

Fig. 5b. Longitudinal variation of anthropogenic surface forcing in the IGB during the winter (WIN), pre-monsoon (PRM), monsoon (MON), and post-monsoon (POM) seasons. The vertical bars through each point represent the standard deviation for 2001–2005.

up along the major urban locations and continued until January. The anthropogenic atmospheric forcing remains high ($>+12\,\mathrm{W\,m^{-2}}$) over the major urban locations throughout the year. However, during the winter due to the low-level inversions, a pool of high anthropogenic atmospheric forcing ($>+24\,\mathrm{W\,m^{-2}}$) exists over the entire IGB. The highest anthropogenic contribution to the observed aerosol forcing in the IGB comes from the seven mega cities (Fig. 1). The longitudinal variation of anthropogenic surface forcing (Fig. 5(b)) shows reverse trend as that of surface ADRF. The anthropogenic surface forcing is less than $-15\,\mathrm{W\,m^{-2}}$ for most of the places in the western and central IGB, but increases sharply toward the east of that for the winter and pre-monsoon seasons. Anthropogenic surface forcing is least during the pre-monsoon season in the western IGB, east of which, the forcing values for monsoon and post-monsoon seasons catch up with each other and the pre-monsoon forcing dips below $-15\,\mathrm{W\,m^{-2}}$ to catch up with the wintertime forcing values. The strong spatial heterogeneity in anthropogenic contribution to ADRF in the IGB is due to complex nature of aerosol loading in the region, resulting from the mixing of anthropogenic and natural aerosols. Past studies (Ref. 5 and the references therein) have shown that the natural dusts are transported every year to the IGB, but their effect on the ADRF is dominant mostly in the western and central parts of the basin. The eastern part remains relatively unaffected from the dust storms.

3.3. *Implications to regional climate*

The anthropogenic fraction is responsible for 67.6% to the heating rate observed over Kanpur region. Over the IGB, the mean annual heating rate is $0.64 \pm 0.19\,\mathrm{K\,day^{-1}}$, of which 54.7% contribution comes from anthropogenic aerosols. The annual heating rate surrounding the major urban locations in IGB is listed in Table 1. It reveals two points; firstly, similar value across the locations implies that the pollution level due to anthropogenic activities leading to the formation of haze is similar in major urban areas. Secondly, the role of anthropogenic components is much more dominated in the eastern IGB, as the dusts are not transported up to that much distance, as discussed earlier. The non-absorbing components contribute more to the TOA forcing, whereas the absorbing components contribute more to surface forcing. Hence, when both are present in abundance, the atmospheric heating would be more, as in the case of winter when BC is the major absorbing component and in summer when pure dust and dust–BC mixture are the absorbing components. The most important implication for such high negative surface forcing is the persistent large reduction of direct solar radiation in the IGB. On an average, aerosols reduce the net solar radiation at the surface over the IGB by 14% on the annual scale, which is even higher (18–20%) over polluted urban locations. The INDOEX observations have shown that the large surface cooling and atmospheric heating affect the regional hydrological cycle.[2] As most of the aerosols are concentrated in the lower atmosphere, that too within the first few km, atmospheric heating is most effective in this region. The enhanced heating due to the mixing of anthropogenic and natural aerosols in this region needs to be addressed in future.

Table 1. Mean (±SD) annual atmospheric heating rate and anthropogenic contribution to Indo-Gangetic haze over major urban locations.

Cities	Locations	Heating rate in $\mathrm{K\,day^{-1}}$	Anthropogenic contribution (in %)
Delhi	28.38N, 77.17E	0.86 ± 0.28	67.9
Agra	27.17N, 77.58E	0.87 ± 0.3	57.3
Kanpur	26.28N, 80.20E	0.84 ± 0.27	65.6
Allahabad	25.25N, 81.58E	0.8 ± 0.28	66.5
Varanasi	25.22N, 83E	0.84 ± 0.28	62.8
Patna	25.35N, 85.12E	0.82 ± 0.23	76
Kolkata	22.36N, 88.24E	0.72 ± 0.16	76.8

4. Summary

In this study, we assessed the anthropogenic contribution to the ADRF in the IGB and studied its spatial and temporal distribution. The Indo-Gangetic haze induces mean annual forcing of -2.9 ± 4.3, -25.5 ± 12.9, and $+22.6 \pm 11.9\,\mathrm{W\,m^{-2}}$ in clear-sky condition. Over Kanpur, the anthropogenic aerosols account for more than 76% of the total aerosol mass, and contribute to 60–65% of the total ADRF at surface annually. On regional scale, the contributions of anthropogenic (~51%) and natural (~49%) aerosols are almost equal, suggesting the importance of natural aerosols in this region. The surface cooling is highest in the winter and pre-monsoon seasons, whereas the spatial heterogeneity is highest in the pre-monsoon and monsoon seasons. Less spatial heterogeneity in the winter and post-monsoon seasons suggest that the anthropogenic pollution is quite uniform throughout the IGB. Anthropogenic contribution is persistently found to be high in the eastern IGB (>70%). Mean annual clear-sky TOA, surface, and atmospheric forcing due to anthropogenic aerosols in the IGB are $+0 \pm 6.8$, -12.6 ± 6.7, and $+12.6 \pm 6.9\,\mathrm{W\,m^{-2}}$, respectively.

The atmosphere is getting heated by more than $0.9\,\mathrm{K\,day^{-1}}$ in the winter and pre-monsoon seasons. Anthropogenic aerosols contribute 55% to the mean (±SD) annual heating rate of $0.64 \pm 0.19\,\mathrm{K\,day^{-1}}$ over IGB. Persistently, large reduction of surface solar irradiance would affect the radiative balance at the surface by decreasing either the latent heat flux through suppressed evaporation or sensible heat flux or both. Such large aerosol loading in the atmosphere would also alter the cloud microphysics through indirect effect. Hence there is a need to understand the effect of the aerosols on the regional hydrological cycle in much better way, which would require more *in situ* measurements to reduce the present-day uncertainty of the ADRF.

Acknowledgments

The authors acknowledge the support of Department of Science and Technology. The efforts of PIs of Kanpur AERONET site are appreciated.

References

1. J. T. Houghton *et al.*, *Climate Change 2001: The Scientific Basis* (Cambridge University Press, 2001), 881 pp.
2. V. Ramanathan *et al.*, *J. Geophys. Res.* **28** (2001) 28371.

3. Y. J. Kaufman, O. Boucher, D. Tanré, M. Chin, L. A. Remer and T. Takemura, *Geophys. Res. Lett.* **32** (2005) doi:10.1029/2005GL025123.
4. C. E. Chung, V. Ramanathan, D. Kim and I. A. Podgorny, *J. Geophys. Res.* **110** (2005) doi:10.1029/2005JD006356.
5. S. Dey, S. N. Tripathi, R. P. Singh and B. N. Holben, *J. Geophys. Res.* **109** (2004) doi:10.1029/2004JD004924.
6. R. P. Singh, S. Dey, S. N. Tripathi, V. Tare and B. N. Holben, *J. Geophys. Res.* **109** (2004) doi:10.1029/2004JD004966.
7. N. Chinnam, S. Dey, S. N. Tripathi and M. Sharma, *Geophys. Res. Lett.* **33** (2006) doi: 10.1029/2005GL025278.
8. V. Ramanathan and M. V. Ramana, *Pure Appl. Geophys.* **162** (2005) 1609.
9. S. N. Tripathi, S. Dey, V. Tare and S. K. Satheesh, *Geophys. Res. Lett.* **32** (2005) doi:10.1029/2005G022515.
10. S. N. Tripathi, S. Dey, V. Tare, S. K. Satheesh, S. Lal and S. Venkataramani, *Geophys. Res. Lett.* **32** (2005) doi:10.1029/2005G022564.
11. S. N. Tripathi *et al.*, *J. Geophys. Res.* **111** (2006) doi:10.1029/2006JD007278.
12. S. Dey and S. N. Tripathi, *J. Geophys. Res.* **112** (2007) doi:10.1029/2006007267.
13. V. Tare *et al.*, *J. Geophys. Res.* **111** (2006) doi:10.1029/2006JD007279.
14. D. Ganguly, A. Jayaraman, T. A. Rajesh and H. Gadhavi, *J. Geophys. Res.* **111** (2006) doi:10.1029/2005JD007029.
15. S. Ramachandran, R. Rengarajan, A. Jayaraman, M. M. Sarin and S. K. Das, *J. Geophys. Res.* **111** (2006) doi:10.1029/2006JD007142.
16. P. Pant *et al.*, *J. Geophys. Res.* **111** (2006) doi:10.1029/2005JD006768.
17. S. Singh, S. Nath, R. Kohli and R. Singh, *Geophys. Res. Lett.* **32** (2005) doi:10.1029/2005GL023062.
18. B. N. Holben *et al.*, *Rem. Sens. Environ.* **66** (1998) 1.
19. K. K. Moorthy *et al.*, ISRO-GBP Sci. Rep. 03-99 (1999).
20. S. Dey, S. N. Tripathi, R. P. Singh and B. N. Holben, *Atmos. Env.* **40** (2006) 445.
21. S. N. Tripathi, A. K. Srivastava, S. Dey, S. K. Satheesh and K. K. Moorthy, *Atmos. Environ.* in press.
22. P. Ricchiazzi, S. Yang, C. Gautier and D. Sowle, *Bull. Amer. Meteor. Soc.* **79** (1998) 2101.
23. S. Dey and S. N. Tripathi, *J. Geophys. Res.*, under review, 2007.

Advances in Geosciences
Vol. 10: Atmospheric Science (2007)
Eds. J. H. Oh and G. P. Singh

PRECISE MEASUREMENT OF POLARIZATION PLANE ROTATION OF PROPAGATING BEAM DUE TO ATMOSPHERIC DISCHARGE

TATSUO SHIINA and TOSHIO HONDA
Graduate School of Advanced Integration Science, Chiba University, 1-33 Yayoi-cho, Inage-ku, Chiba-shi, Chiba 263-8522, Japan

TETSUO FUKUCHI
Central Research Institute of Electric Power Industry, 2-6-1 Nagasaka, Yokosuka-shi, Kanagawa 240-0196, Japan

The rotation of polarization plane of a propagating beam caused by a high voltage discharge was experimentally examined. Using a CW visible laser and repeating mirror optics, the rotation angle of <1 deg was measured with a differential detection of >30 dB. The waveform of the rotation angle showed good correlation with the discharge current waveform, with a correlation coefficient of >0.94.

1. Introduction

Meteorological observation technology has sufficiently developed to enable accurate weather prediction, while local disasters such as heavy rain and lightning stroke have not decreased. In Japan, flood and lightning cause over 10 and 100 deaths per year, respectively. Automated Meteorological Data Acquisition System (AMeDAS), which is operated by the Japan Meteorological Agency, uses a cell size of 20 km square area. Observations using meteorological satellites use a cell size of 4 km square area. Therefore, it is difficult to detect rapid local weather change, which accompany heavy rain and lightning strikes.

Electromagnetic antennas are mainly used to detect lightning strikes. Lightning detection and positioning systems such as SAFIAR and LPATS are in operation in Japan. The antenna measurement requires numerous observation points synchronized in time to obtain the lightning strike position, and lacks flexibility of measurement. Radar has the flexibility of single-point measurement, while its use is limited because of legal restrictions on the use of electromagnetic waves, and its measurement resolution may be insufficient. Therefore, a highly maneuverable measurement method, which can follow the local weather change, is desired. A lidar using the Stark effect has been proposed,[1] although it has not yet achieved its ultimate goal.

As a tool for prediction of heavy rain and lightning strikes and prevention of local disasters resulting from such phenomena, an in-line type micropulse lidar (MPL) system has been developed.[2–4] This system can distinguish ice crystals from water particles, and the flow of ice crystals can be used to derive meteorological parameters linked to heavy rain and lightning strikes. The MPL system uses in-line optics, which allows a constant overlap of the transmitting and receiving field-of-views (FOVs), and can receive the lidar return signal with no obscuration distance (i.e. the return signal can be obtained from the immediate front of the system) by using an outgoing annular beam. Furthermore, an optical circulator is used to separate the transmitted beam and the return signal. The lidar return signal of mutually perpendicular polarizations (parallel and perpendicular to the polarization of the outgoing beam) can be separately detected, which allows precise measurement of the depolarization ratio. The FOV of 0.1 mrad is sufficiently narrow to eliminate the depolarization effect caused by multiple scattering, so the lidar system can distinguish ice crystals from depolarization. The MPL system aims to derive the metrological parameters connected to heavy rain and lightning strikes by monitoring the movement of ice crystals, and is currently in year-round observation.

However, the movement of ice crystals is only indirectly associated with local disasters such as heavy rain and lightning strikes. Furthermore, the effects of seasonal change, regional difference, and variations in the air current must be removed. If the metrological parameters connected to lightning strikes could be obtained more directly, accurate prediction would become possible.

The authors focused attention on the magneto-optical effect (Faraday effect). In a partially ionized atmosphere, a propagating beam experiences a rotation in the polarization plane due to the electromagnetic pulse accompanying a lightning strike. In this study, the feasibility of optical remote measurement of the change in the electromagnetic field or in the ionized density distribution in a high-voltage discharge is examined. By a laboratory-based experiment using an impulse voltage discharge, the polarization extinction ratio and the receiver's dynamic range for the detection of the polarization rotation angle are examined. The experimental results are compared with the results of analytical calculations.

2. Principle — Faraday Effect

The interaction between the light and the atmosphere is caused by scattering (Mie, Rayleigh, Raman), and also by the magneto-optical effect and electro-optical effect. The magneto-optical effect (Faraday effect) is associated with the lightning discharge. It has been reported as an optical measurement method for magnetic confinement fusion reactor.[5] The polarization plane of a beam propagating parallel to the magnetic flux is rotated in a partially ionized atmosphere (plasma) (Fig. 1). The rotation angle is proportional to the product of the ionization electron density n_e and the magnetic flux density B along the beam propagation path. The linearly polarized beam can be regarded as a combination of the clockwise and the counterclockwise circularly polarized beams. The refractive indices of the ionized atmosphere for each circularly polarized beam are as follows:

$$n_{\pm} = \left(1 - \frac{\omega_{\mathrm{pe}}^2}{\omega^2}\frac{\omega}{\omega \pm \omega_{\mathrm{ce}}}\right)^{1/2},$$

$$\omega_{\mathrm{pe}} = \sqrt{\frac{e^2 n_{\mathrm{e}}}{\varepsilon_0 m_{\mathrm{e}}}} \quad \omega_{\mathrm{ce}} = \frac{eB}{m_{\mathrm{e}}} \tag{1}$$

where ω_{pe}, ω_{ce} are the plasma and electron cyclotron frequencies, respectively, e is the fundamental charge, m_{e} is the electron mass, and ε_0 is the permittivity of free space. Therefore, the rotation angle of polarization of

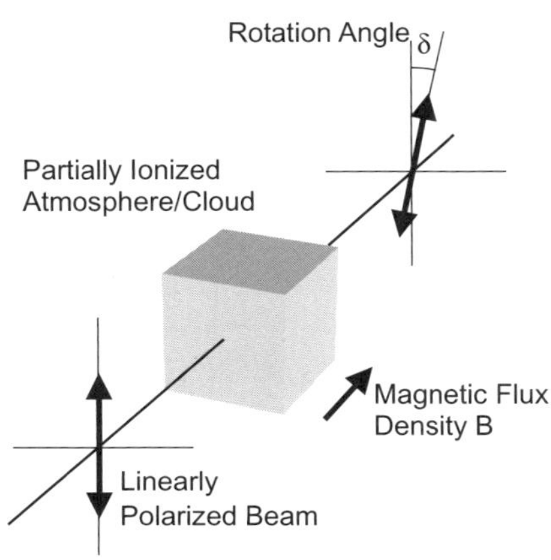

Fig. 1. Faraday effect.

the beam propagated at distance $L(= L_1 - L_2)$ is obtained as follows:

$$\delta = \frac{\pi}{\lambda}\int_{L_1}^{L_2}(n_+ - n_-)dl$$
$$= 2.62 \times 10^{-13}\lambda^2 \int_{L_1}^{L_2} n_e B dl, \qquad (2)$$

where λ is the wavelength of the propagating beam. Since δ is proportional to λ^2, the rotation angle for visible light is small. Therefore, the polarization angle rotation must be measured with high accuracy in order to detect lightning discharge.

When the Faraday effect is applied to lightning measurement, the atmosphere needs to be partially ionized, and the magnetic flux due to the lightning discharge must exist. Cloud-to-cloud discharge, which causes 20–30 times continuous discharge, satisfies those conditions.

3. High-voltage Discharge Experiment

3.1. *Experimental model*

An experimental apparatus incorporating multiple reflection optics was constructed to verify whether the polarization plane rotation of a propagating beam due to an electrical discharge can be detected. The experimental setup is shown in Fig. 2, and the specifications are shown in Table 1. The propagating beam runs repeatedly around the discharge path so that the beam can interact with the high-voltage discharge multiple times, providing enough rotation in the polarization plane due to the Faraday effect. The optics consists of input–output optics and a square mirror. The latter is installed inside a discharge chamber.

The discharge gap consisting of two needle electrodes is at the center of the square mirror. The polarization and divergence of the beam are adjusted on entering the square mirror. The total length of the optical path can be changed by controlling the tilts of the four sides of the square mirror.

The outgoing beam from the square mirror is divided by a polarized beam splitter and detected separately at mutually orthogonal polarizations. The polarization of the incident beam is adjusted to balance the detected intensities. An impulse voltage is applied to the needle electrode, causing a spark discharge inside the chamber. This causes a rotation in the polarization plane of the propagating beam, and the difference between the intensities of the two mutually orthogonal polarizations is detected by a differential amplifier.

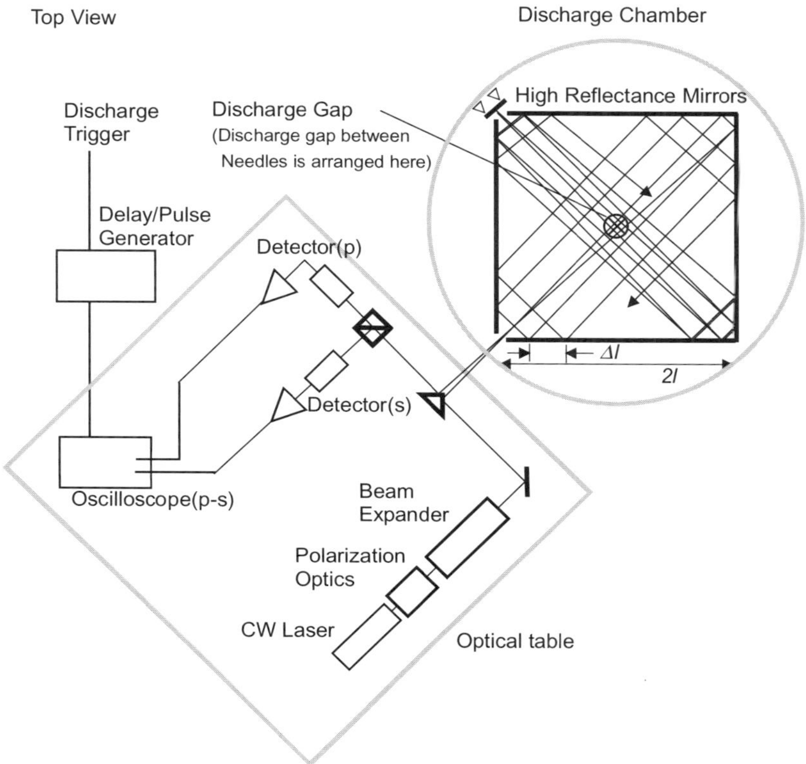

Fig. 2. High-voltage discharge experiment.

Table 1. Specification of high-voltage discharge experiment.

Light source	YAG Laser ($\lambda = 532\,\mathrm{nm}$) 150 mW (CW)
Detection	PDs with Amp. + Differential Amplifier
Discharge gap	3–10 cm
Discharge voltage	180 kV (max)
Beam path length	19–50 m (max)

3.2. *Simulation*

A numerical calculation of the polarization plane rotation was performed. The discharge current waveform was defined as that of the lightning return stroke (peak current: 2 kA, 1/10 of the typical return current of cloud-to-ground discharge).[6–8] The ionized electron density is assumed that the atmospheric molecules were perfectly ionized as 100% ($\cong 10^{25}/m^3$) around the discharge path of 2 cm diameter and partially ionized ($<10^{15}/m^3$) in other areas. The magnetic flux density was calculated with the considerations of the distance from the discharge path and its orientation.

Figure 3 shows an example of calculated results for the experimental model. The sides of the square mirror were 28 cm in length. The beam re–flection step was 2 cm. The propagating beam was reflected multiple times by the four sides of the square mirror, as shown in Fig. 3(a). Mirror M' returns the propagating beam to the output from the square mirror.

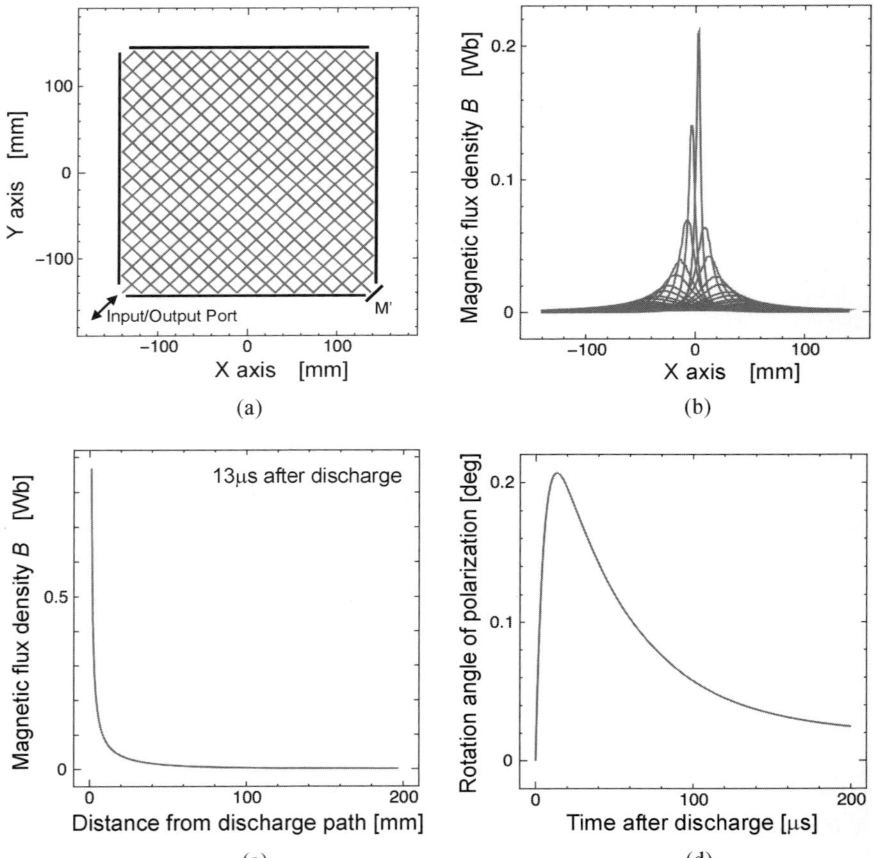

Fig. 3. Calculation result. (a) Optical path, (b) Magnetic flux density B, (c) B vs distance and (d) Rotation angle.

Table 2. Estimation of rotation angle of beam polarization under various conditions.

Mirror length (mm)	Number of reflections/ step (mm)	Necessary reflectance (%)	Total optical path length (m)	Discharge gap (cm)	Rotation angle (max) (deg.)	Polarization ratio
280	12/21.4	91	19	10	0.21	1:1.007 (21 dB)
280	28/10.0	96	44	10	0.51	1:1.018 (18 dB)
280	56/05.0	98	89	10	0.95	1:1.034 (15 dB)
500	12/41.7	91	34	10	0.074	1:1.003 (26 dB)
500	50/10.0	98.3	141	10	0.51	1:1.018 (18 dB)
500	12/41.7	91	34	20	0.074	1:1.003 (26 dB)
500	50/10.0	98.3	141	20	0.52	1:1.018 (17 dB)

The magnetic flux densities along the beam propagation projected on a side of the square mirror are shown in Fig. 3(b). Multiplying the magnetic flux density B by the electron density n_e along the propagation distance, the rotation angle of the beam polarization was estimated by Eq. (2). Figure 3(c) shows the change of the magnetic flux density with respect to the distance from the discharge path at 13 μs after the discharge. The flux density changes its value along the elapsed time. The magnetic flux density B is inversely proportional to the distance from the discharge path.

The electron density n_e is assumed to be high only within 2 cm of the discharge path. Based on the results of Fig. 3(a)–(c), the rotation angle of the polarization was estimated, and is shown in Fig. 3(d). The change of the rotation angle had the same time response as the discharge current. The result shows that the frequency response of a few MHz was sufficient for the detector and amplifiers. Other calculations with different discharge gaps and the beam propagation conditions are summarized in Table 2. Selecting those conditions, we can distinguish the rotation angle of the beam polarization by differential detection with a dynamic range of 30 dB.

4. Experiment

4.1. *Experimental setup*

The experimental setup of Fig. 2 was constructed and installed. The specification of experimental apparatus is the same as Table 1. Figure 4 shows photographs of the optics (left) and the square mirror in the discharge chamber (right). The sides of the square mirror were 28 cm in length, and its reflectance was about 95%. The number of the reflections on a single mirror was fixed at 12–14. The total round-trip optical path length was

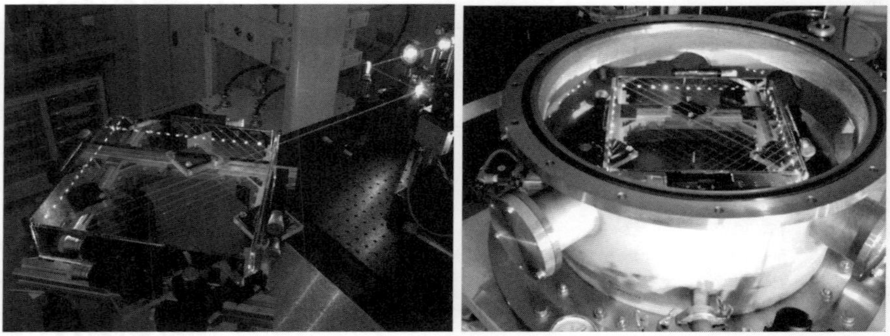

Fig. 4. Experimental optics and square mirror.

about 20 m. The incident laser beam power of 150 mW was attenuated to less than 100 μW at the output because of the reflection loss of the mirrors and divergence of the beam. The discharge gap was placed at the center of the chamber. The discharge gap length was 0–10 cm.

The rotation angle of the beam polarization was estimated as 0.21 degrees (max) in the above situation, so the polarization ratio will become 1:1.007. It means that the differential detection of the dynamic range of 30 dB enables to distinguish the difference.

4.2. *Experimental result*

Waveforms of the differential amplifier output, which represent the rotation of the polarization plane, are shown in Fig. 5. The charge voltage was 50 kV, and the discharge gap between the needles was 5 cm. Figure 5(a) shows the output for 40 consecutive discharge shots. Since the discharge path varied for each shot, the relative alignment between the discharge path and the propagating beam was changed. Waveforms which showed similarity with the discharge current waveform are indicated as a black solid curve. Although the output signals were triggered by the charge voltage, the black solid curves had a delay time of about 25 μs in average. We confirmed that the delay was caused by the amplifier circuits and cables.

One of the black solid curves is enlarged in Fig. 5(b). The output of the differential amplifier when the incident beam was interrupted in front of the detectors is also shown. This shows that the electromagnetic noise from the discharge did not influence the output signal. The correlation coefficient between the black solid curves and the discharge current waveform

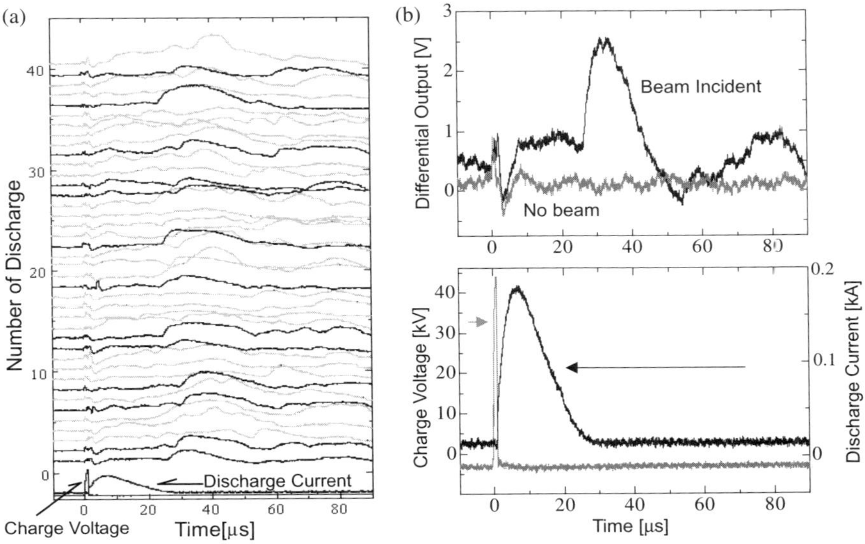

Fig. 5. Experimental result of rotation angle of propagating beam's polarization plane. (a) Differential outputs of 40 times discharge event, (b) validation of differential output and discharge current.

was >0.94. This result was in agreement with the result of the simulation, which showed that the rotation of the polarization plane had the same time response as the discharge current.

The rotation angle of the polarization plane was estimated from the measured results. Considering the amplification of the differential amplifier (×100), the intensity ratio of the mutually orthogonal polarization outputs was 1.00:0.98. Therefore, the rotation angle δ is calculated as

$$\delta = \frac{\pi}{4} - \tan^{-1}(\sqrt{0.98}/\sqrt{1.00}) = \tan^{-1}\left(\frac{\sqrt{1.00}-\sqrt{0.98}}{\sqrt{1.00}+\sqrt{0.98}}\right) = 0.29\ (\text{deg.})$$

This value also agreed with the result of the simulation.

The correspondence between the obtained signal with the discharge current showed that the propagating beam interacted with the discharge only at a single optical path. Larger rotation angle could be obtained by increasing the discharge current and gap length. Since the discharge chamber used in this study can only accommodate discharge gaps of about 10 cm, the same experiment will be performed in open air, using discharge gaps in the order of 1 m and a larger impulse voltage generator.

5. Summary

The rotation of the polarization plane of a beam propagating through a region of high-voltage discharge was estimated. To extend the optical path length in the vicinity of the discharge, repeating mirror optics was devised. The experimental setup was installed in a discharge chamber. As a result, the rotation angle was estimated as <1 deg. To detect such a small angle, the sensitivity of the system needed to have a dynamic range of 30 dB. The rotation of the polarization plane was experimentally measured to be 0.29 deg, which was in agreement with the simulation results.

Electromagnetic measurement using the Faraday effect can be applied to lidar measurement of lightning discharges. As the pulse beam enables time-resolved detection, evaluation of the lightning strike position, measurement of the change in the electromagnetic field, and spatial distribution of ionization should be possible.

References

1. V. Gavrilenko, K. Muraoka and M. Maeda, Proposals for remote sensing of electric field under thunderclouds using laser spectroscopy, *Jpn. J. Appl. Phys.* **39** (2000) 6455–6458.
2. T. Shiina, K. Yoshida, M. Ito and Y. Okamura, In-line type micro pulse lidar with annular beam — Experiment, *Appl. Opt.* **44**(34) (2005) 7407–7413.
3. T. Shiina, K. Yoshida, M. Ito and Y. Okamura, In-line type micro pulse lidar with annular beam — Theoretical approach, *Appl. Opt.* **44**(34) (2005) 7467–7473.
4. T. Shiina, E. Minami, M. Ito and Y. Okamura, Optical circulator for in-line type compact lidar, *Appl. Opt.* **41**(19) (2002) 3900–3905.
5. F. Simonet, Measurement of electron density profile by microwave reflectometry on tokamaks, *Rev. Scientific Instrum.* **56** (1985) 664–669.
6. Y. T. Lin, M. A. Uman and R. B. Standler, Lightning return stroke models, *J. Geophys. Res.* **85**(C3) (1980) 1571–1583.
7. M. J. Master, M. A. Uman, Y. T. Lin and R. B. Standler, Calculations of lightning return stroke electric and magnetic fields above ground, *J. Geophys. Res.* **86**(C12) (1981) 12.127–12.132.
8. M. A. Uman, Lightning return stroke electric and magnetic fields, *J. Geophys. Res.* **90**(D4) (1985) 6121–6130.

Advances in Geosciences
Vol. 10: Atmospheric Science (2007)
Eds. J. H. Oh and G. P. Singh

CHARACTERISTICS FOR OPTICAL PROPERTIES OF BACKGROUND AEROSOLS, WATER, AND DUST CLOUDS MEASURED BY USING LIDAR OVER CHUNG-LI, TAIWAN

C. W. CHIANG[*,†,‡], S. K. DAS[†] and J. B. NEE[†]
[*] *Research Center for Environmental Changes, Academia Sinica, Nankang, Taiwan*

[†] *Department of Physics, National Central University, Chungli, Taiwan*
[‡] *cwchiang@alumni.ncu.edu.tw*

Aerosol and cloud studies are important as they affect both the radiation and the climate of the atmosphere. In this paper, we present and discuss the correlation between the optical properties of aerosol and cloud (<3 km above ground level) over Chung-Li (24.58°N, 121.10°E). This study was carried out by using ground-based lidar and *in situ* balloon measurements from 2002–2004. Results show that there are differences in the properties of the aerosol and cloud measurements in the spring season (March–May), which may be due to the differences in transport sources and aerosol compositions. These facts are supported by backward trajectories, radiosonde, and height distributions of aerosol-cloud backscatter and depolarization ratio. For background aerosols, the depolarization ratio shows a decreasing trend with increasing relative humidity and the backward trajectories displayed the origin of air masses concluding that the source is from the coastal industrial area of China. Dust-clouds traced back from Northeast China in Spring are marked by higher depolarization and backscatter ratios, which is also sensitive to relative humidity. The statistical properties of water-clouds and dust-clouds were also accounted for.

1. Introduction

Both aerosols and clouds, due to their influences on the environment and climate are always interesting aspects for various purposes of atmospheric studies. Aerosols have the potential to change the global climate because of their direct radiative effects and indirect effects as cloud condensation

[‡]Corresponding author.

nuclei (CCN). Radiative and microphysical properties of aerosols are variable in time and space. The vertical distribution of aerosols is an important parameter to be measured, in order to evaluate their radiative effect.[1,2] In addition, the relationship between the aerosol properties and ambient relative humidity (*RH*) is also important for atmospheric studies.[3] In particular, there is little information about the vertical distribution of aerosol properties in the atmosphere due to the technical difficulties and analytical limitations. In addition, there are few observational data on the relation between the aerosol properties and ambient *RH*. Since the aerosol shape, size, and chemical composition critically depend on *RH*, the properties of atmospheric aerosol particles can be expressed as function of the *RH* at thermodynamic equilibrium with the surrounding atmosphere.[4]

The remote sensing depolarization techniques are developed for the above-mentioned requirements. A depolarization lidar can provide information on particle shape and size by employing the short pulses of polarized light of laser. The radiation backscattered by homogeneous spherical particles will maintain the original (parallel) polarization, whereas nonspherical particles will induce some degree of depolarization. Thus, the degree of depolarization will provide the distribution of particle and water in different phases. Lidar polarization measurements for many cloud states existing below 3 km have nonzero depolarization ratios (*DR*), but these clouds should be liquid clouds consisting of spherical droplets. These liquid clouds consisting of aspherical particles appeared mainly in the spring season during which Asian dust mainly burst. In this chapter, we have studied and discussed the optical properties of aerosols and clouds in the altitude range of 0.7–3 km by using depolarization lidar during 2002–2004. The depolarization ratios for different sources of clouds are traced by HYSPLIT (Hybrid Single-Particle Lagrangian Integrated Trajectory) calculations. The statistics of the optical properties of aerosols and clouds with respect to relative humidity have also been discussed.

2. Lidar Methodology and Related Definitions

2.1. *Lidar system*

The lidar system used in the present study consisted of a Nd:YAG laser at a wavelength of 532 nm. The receiver consists of a 20 cm Schmidt–Cassegrain telescope. Signals are measured by photomultiplier tubes (PMT) and analyzed by multi-channel analyzers (MCA) with a vertical resolution

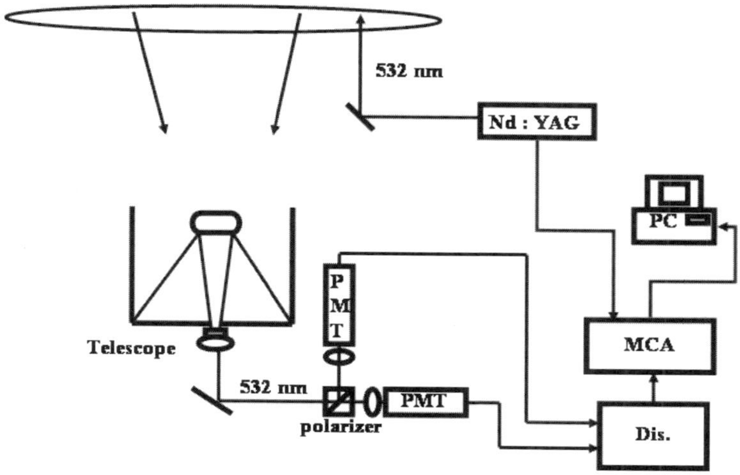

Fig. 1. Block diagram of lidar system at Chung-Li, Taiwan. (PMT: photomultiplier tube, Dis: discriminator, MCA: multi-channel analyzer, PC: personal computer).

of 24 m and a temporal resolution of 33 s. The schematic diagram of lidar system is shown in Fig. 1. Details about our lidar studies/system can be found in the studies of Nee *et al.*[5] and Chen *et al.*[6] The lidar measures height profile of backscattered signal from aerosols, which are converted into backscattering ratio. The aerosols backscattering ratio, *BR*, is defined as follows:

$$BR(z) = \frac{(\beta_a(z) + \beta_m(z))}{\beta_m(z)}, \tag{1}$$

where β_a and β_m are the backscatter coefficients of aerosols and molecules, respectively. The backscattering coefficient of molecule is derived from the molecular concentration measured by radiosonde multiplied by total Rayleigh scattering cross section. Calculation of aerosol backscattering ratio involved Fernald solution.[7] For this study, the initial condition for the upper height is set at 5 km, and the reference value of backscattering ratio is 1.05, based on the earlier study.[8] The extinction-to-backscatter ratio or lidar ratio for the aerosol and cloud cases was derived by Chiang *et al.*[9]

The polarization measurements were carried out by recording the lidar returns for perpendicular and parallel polarizations. The aerosol depolarization ratio (δ_a) is calculated as

$$\delta_a(z) = \frac{\delta(z) * BR(z) - \delta_m}{BR(z) - 1}, \tag{2}$$

where δ is the total (aerosol + molecule) depolarization ratio which is obtained by taking the ratio of the perpendicular and the parallel intensity relative to the outgoing laser beam. δ_m is the molecular depolarization ratio ($\delta_m \sim 1.4\%$).[10] Equation (2) also indicates how correlation of the above two traces namely, BR and δ_a, along a vertical interval Δz will provide the information of solid or liquid particles' dominance. The depolarization ratio is 0 for spherical particles and deviates from 0 for nonspherical particles such as mineral dust.

2.2. *Background aerosol and cloud definitions*

In order to understand the optical properties of cloud in the height region of 0.7–3 km, we need to distinguish the differences between the clouds and aerosols. We defined the background aerosols (BGA) based on their scattering properties which follows two criteria: (i) the backscattering coefficient should be smaller than five times that scattered by air molecules and (ii) the DR of aerosols should be smaller than the value of 0.1 at the aerosol layer (Fig. 2). BGA normally consists of pollutants, sea salt,

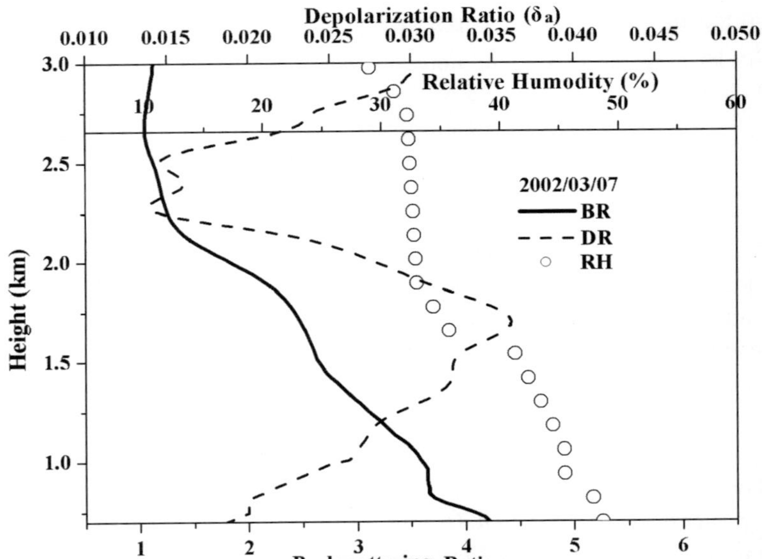

Fig. 2. The height distribution of backscattering ratio (—), depolarization ratio (---), and relative humidity (○) on 7 March 2002.

and other unknown particles. Dust is a kind of aerosol that only occurs occasionally in the spring time. So, they should be treated as separate cases in addition to BGA.

For clouds definition we have to consider two points: First, the backscattering coefficient should be five times greater than that scattered by air molecules. Such a higher backscattering coefficient usually occurs when the scattering is by larger particles (such as water drop), and it is independent of the measured wavelengths of the laser. Second, the local *RH* should be higher than 70% at the cloud height. Usually, the clouds observed in the regions (0.7–3 km) have temperature higher than 0°C with *DR* close to 0, because they consist of water drops (refer to Fig. 5), except for some special events such as the occurrence of dust storm from Northern China (Fig. 9).

3. Results and Discussion

3.1. *Optical properties of BGA and water clouds*

Figure 2 shows the results of *BR*, *DR*, and *RH* for one of the BGA cases on 7 March 2002. The mean height profile of *BR* for 39 cases of BGA in Spring during the period 2002–2004 is shown in Fig. 3. Each case used in the analysis has been integrated over an hour from 20.00 to 21.00 local time, and this time is very close to the time of launching of the radiosonde, and thus both can be compared. We found that most BGA are distributed below 2 km over our location. Figure 4 shows the relationship between the aerosol *DR* and *RH* for all the BGA cases in the altitude range of 0.7–2 km in Spring. On average, the aerosol *DR* of BGA is smaller than 0.06. The mean aerosol *DR* (solid line) shows the decreasing trend with increasing *RH*. The effect of *RH* on aerosol *DR* is expected as the particles are more nonspherical in the dry air. The increased *RH* will reduce the *DR*. This is due to the fact that increased humidity will cause hygroscopic aerosols to grow and its shapes will become more symmetrical like water drops, which will ultimately reduce the *DR*.

Figure 5 shows a water-cloud case on 21 March 2002. The water cloud layers were observed in the height range of 1–2 km. The *DR* of this water cloud shows similar tendency with *BR* in cloud layers. Figure 6 shows the relationship between *DR* and *RH* for 107 water-cloud cases at an altitude range of 0.7–3 km, where no correlation can be identified. The mean *DR* is around 0.015.

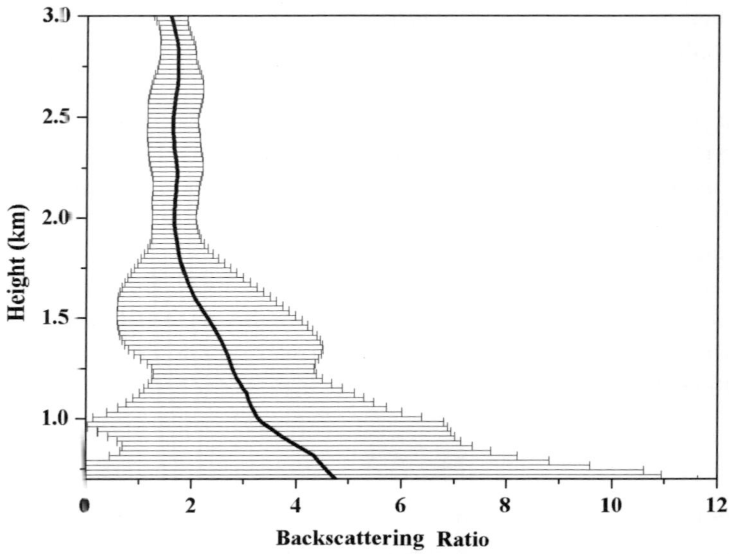

Fig. 3. Backscattering ratio profile of mean background aerosols in spring (March–May) during 2002–2004. The horizontal bars represent the standard deviation from the mean.

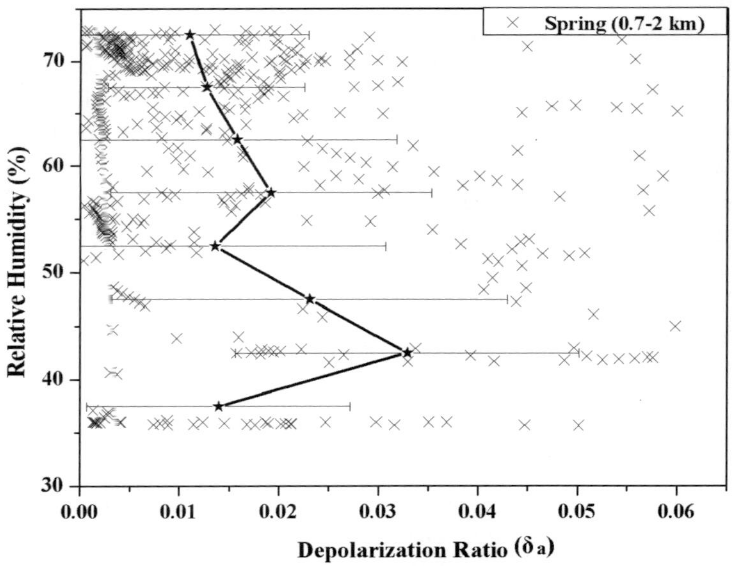

Fig. 4. Depolarization ratio of aerosols (cross symbol) at 0.7–2 km vs RH for all the cases in Spring (March–May) from 2002 to 2004. The solid line shows the mean profile of the DR as a function of RH, obtained by calculating the average (cross symbol) values with interval ranges of 5%. The error bars represent the variability within these intervals of RH ranges.

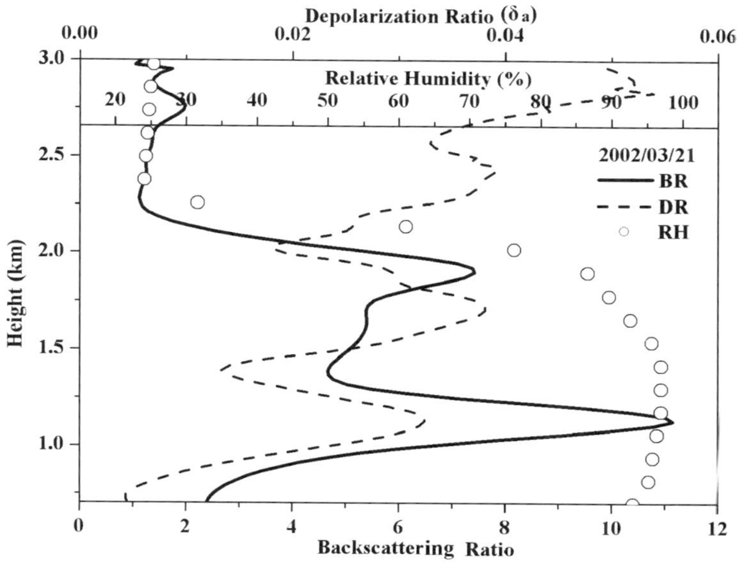

Fig. 5. The height distribution of backscattering ratio (—), depolarization ratio (---), and relative humidity (○) on 21 March 2002.

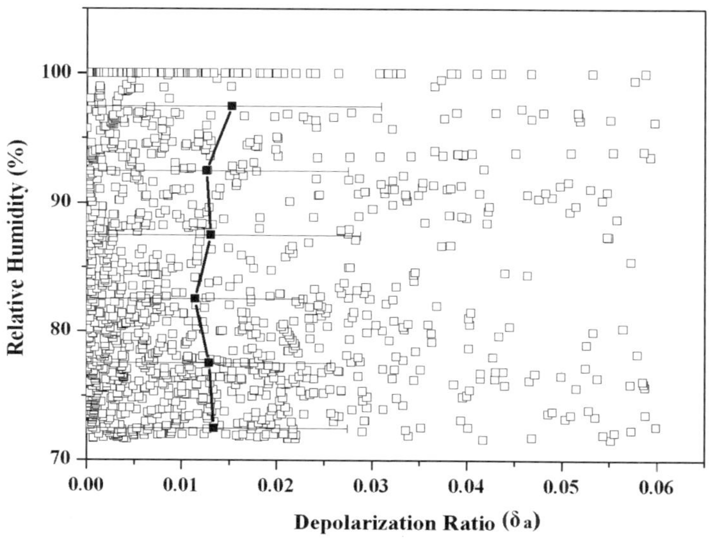

Fig. 6. Depolarization ratio measured for water cloud layers (square symbol) at 0.7–3 km as a function of RH for all the cases in Spring (March–May) from 2002 to 2004. The solid line shows the mean profile of the DR as a function of RH, obtained by calculating the average (square symbol) values with interval ranges of 5%. The error bars represent the variability within these intervals of RH ranges.

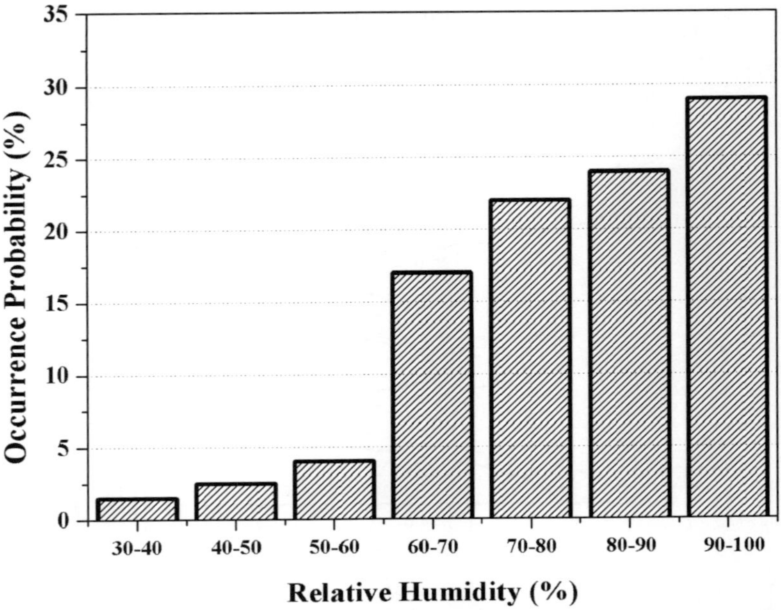

Fig. 7. The occurrence probability of backscattering ratio with the value higher than 6 vs relative humidity.

3.2. *Relationship between BGA and clouds*

The frequency distribution for the probability of occurrence for *BR* higher than 6 is shown in Fig. 7. From these three-year cases, we found that most cases ($BR > 6$) occurred when *RH* was higher than 60%–70%. This can be one of the causative reasons that would lead to the growth of the aerosol particles by condensation, thus changing their optical properties. This result is similar to that of Hanel[3] and Fitzerald *et al.*[11]

The correlation of maximum extinction coefficient and maximum *DR* of cloud cases is shown in Fig. 8. Most of the cloud layers ($RH > 70\%$) have the maximum *DR* lower than 0.1. The cases, which have the maximum extinction coefficient higher than 1 (1/km), showed a sharp increasing trend with increasing *DR*. This may be possibly due to the multi-scattering effect or also can be due to the presence of large particles with nonspherical shapes that can produce large scattering which can be related to long-range transport of Asian dust storms with moisture and pollutants.[12]

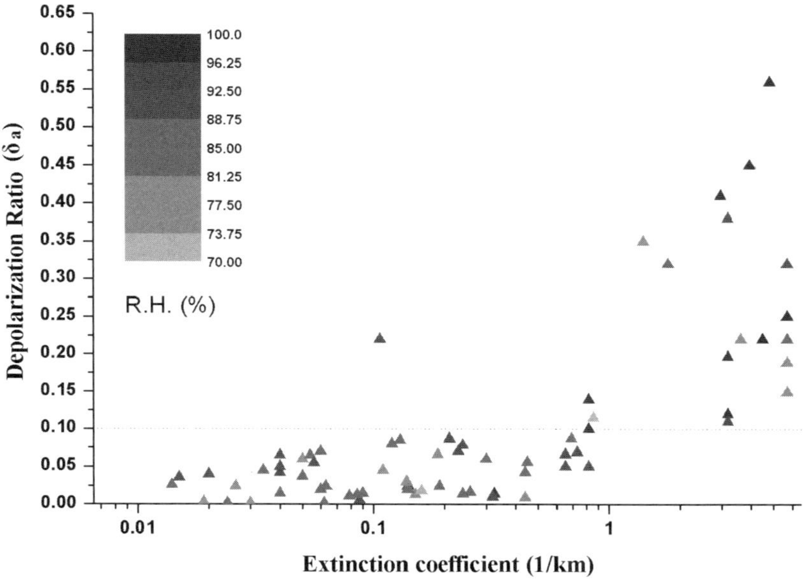

Fig. 8. The relationship between maximum depolarization ratio and maximum extinction coefficient of cloud events with respect to *RH* in Spring from 2002 to 2004.

3.3. *Characteristics of dust clouds*

Our observation location was affected frequently by outside aerosols such as dust storms from Northern China which mostly appear during the spring seasons.[13–15] The term "dust cloud" has been used instead of using "dust layer" because we have measured higher values of backscattering ratio of the dust as compared with other literature. We think that this phenomenon is related with our location as it is of low latitude, where atmosphere is more humid and the dust particles can easily interact with humidity and pollution.[12] Figure 9 shows one of the dust-cloud case on 31 March 2002, which is observed over our lidar site during the period (spring season) of dust storm which break out in Northern China. The peaks of *BR* of this dust cloud are at 1.8 and 2.1 km with *RH* around 100% and 80%, and depolarization ratios are 0.27 and 0.30, respectively. The upper layer, which is around 2.1 km has lower *BR* (~12.5) with *RH* (~80%), but higher *DR* (0.30). The result shows that the dust clouds, which are measured by lidar during spring seasons, accompanied with BGA and water clouds have higher extinction coefficient and *DR*.

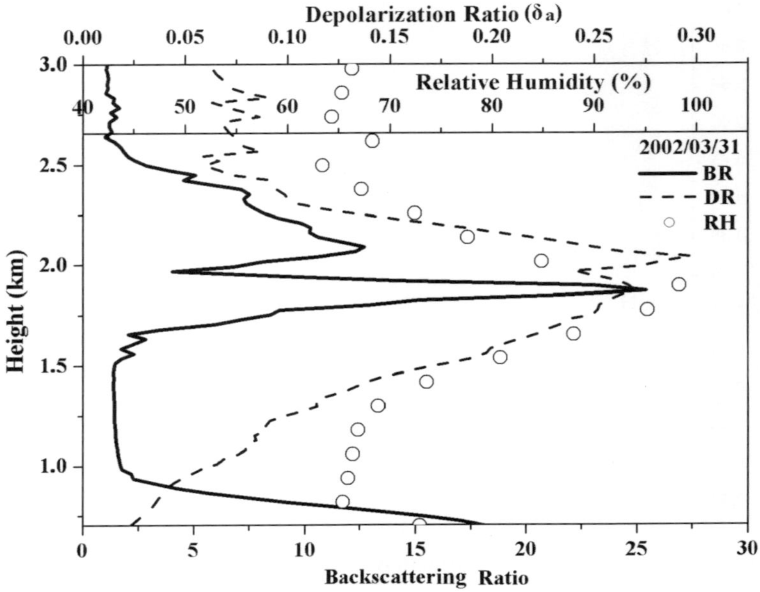

Fig. 9. The height distribution of backscattering ratio (—), depolarization ratio (---), and relative humidity (◯) for a dust case on 31 March 2002.

It is worthy to notice that mineral dust does not absorb water as they were primarily hydrophobic, but from the result (Fig. 9), the *DR* shows the inverse relation with *RH* at 1.8 km, and 2.1 km. The reason is that these dust cloud layers become hydrophilic as dust aerosols are modified during the long-range transport through chemical reactions by mixing with sulfate, sea salt, and other pollutants' activities with ambient humidity to act as giant CCN (cloud condensation nuclei) to form clouds.[16,17]

3.4. *Source-dependent properties*

Figure 10 shows the results from 72 h backward trajectories and displayed the air parcel sources at the altitude of 1 km (Fig. 10(a)) and the sources of all dust cloud layers (nine cases) in Fig. 10(b). These trajectories were generated by HYSPLIT model of NOAA (National Oceanic and Atmospheric Administration). The trajectories of Fig. 10(a) are under cloud-free and dust-free conditions over our site. The times of trajectories are in accordance with the times of BGA cases in 2002–2004. The heights

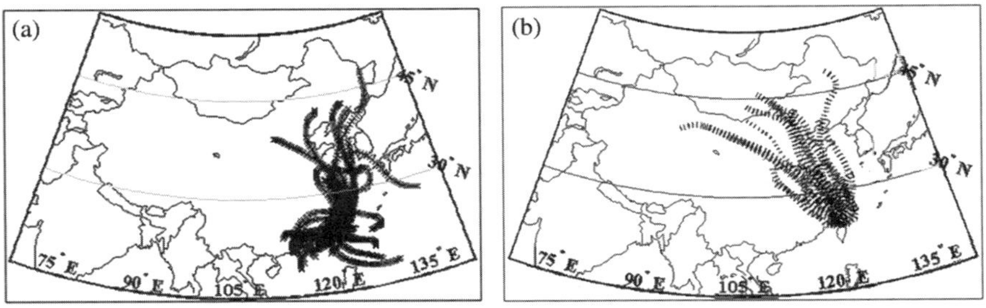

Fig. 10. The results of (a) air parcel sources at altitude 1 km and (b) dust clouds sources from 72 h backward trajectories from the HYSPLIT model.

where *DR* (>0.1) are used for tracing dust-cloud sources. For BGA cases, the air parcels mainly come from the coastal industrial areas of China. The dust-clouds were traced back from the nearby desert regions of Northern China through populated and industrialized areas. Therefore, there were ample opportunities that the dust particles had interaction with moisture and pollutants to form clouds. Those clouds have complex composition and interaction with aerosol, dust, and moisture. The related discussion can be found in the study of Nee *et al.*[12]

4. Summary

In this chapter, the differences in the optical properties of aerosol and cloud are studied based on the depolarization ratio (δ_a) measurements. The polarization lidar observations can provide quantitative information about phase composition of clouds. From the statistics of analysis, the following conclusions can be made:

(1) Most aerosols are distributed below 2 km over our location and their depolarization ratios are smaller than 0.06. The effect of relative humidity on depolarization ratio is expected as the aerosols are more nonspherical in the dry air. The backscattering ratio increases steeply when relative humidity is higher than 60%–70%.
(2) Clouds with higher extinction coefficient and depolarization ratio are caused by dust particles that are mainly transported from Northern China. This is supported by HYSPLIT model. The interactions of aerosols and clouds may be caused by pollutants and humidity.

(3) The occurrence of clouds for air masses coming from Northern China is 36% (for total 117 cases). And the occurrence of dust-clouds with depolarization ratio higher than 0.1 is 21% (among 42 cloud cases), that basically came from the nearby desert region of Northern China in the spring time during the period 2002–2004.

References

1. H. Liao and J. H. Seinfeld, Radiative forcing by mineral dust aerosol: Sensitivity to key variables, *J. Geophys. Res.* **103** (1998) 31637–31645.
2. A. L. Quijano, N. Sokolik and O. B. Toon, Radiative heating rates and direct radiative forcing by mineral dust in cloudy atmospheric conditions, *J. Geophys. Res.* **105** (2000) 12207–12219.
3. G. Hanel, The properties of atmospheric aerosol particles as function of the relative humidity at thermodynamic equilibrium with the surrounding atmosphere, *Adv. Geophys.* **19** (1976) 73–188.
4. R. Dmowska and B. Saltzman, *Advanced Geophysics* (New York Academic Press, 2001).
5. J. B. Nee, C. N. Lin, C. I. Lin and W. N. Chen, Lidar studies of cirrus clouds near the tropopause at Chung-Li, Taiwan (25°N, 121°E), *J. Atmos. Sci.* **55** (1998) 2249–2257.
6. W. N. Chen, C. W. Chiang and J. B. Nee, Lidar ratio and depolarization ratio for cirrus clouds, *Appl. Opt.* **41** (2002) 6470–6476.
7. F. G. Fernald, B. M. Herman and J. A. Reagan, Determination of aerosol height distribution by lidar, *J. Appl. Meteorol.* **11** (1972) 482–489.
8. C. W. Chiang, W. N. Chen, W. A. Liang, S. K. Das and J. B. Nee, Optical properties of tropospheric aerosols based on measurements of Lidar, sun-photometer, and visibility at Chung-Li (25°N, 121°E), *Atmos. Environ.* **41** (2007) 4128–4137.
9. C. W. Chiang, S. K. Das and J. B. Nee, An iterative calculation to derive extinction-to-backscatter ratio based on lidar measurement, *J. Quant. Spectrosc. Radiat. Transfer* **109** (2008) 1187–1195.
10. A. Weber, S. P. S. Porto, L. E. Cheesman and J. J. Barrett, High-resolution Raman spectroscopy of gases with cw-laser excitation, *J. Opt. Soc. Am.* **57** (1967) 19–28.
11. J. W. Fitzerald, W. Hoppel and M. Vietti, The size and scattering coefficient of urban aerosol particles at Washington DC as a function of relative humidity, *J. Atmos. Sci.* **39** (1982) 1838–1852.
12. J. B. Nee, C. W. Chiang, H. L. Hu, S. X. Hu and J. Y. Yu, Lidar measurements of Asian dust storms and dust cloud interactions, *J. Geophys. Res.* **112** (2007) D15202, doi:10.1029/2007JD008476.
13. T. H. Lin, Long-range transport of yellow sand to Taiwan in Spring 2000: Observed evidence and simulation, *Atmos. Environ.* **35** (2001) 5873–5882.

14. C. Y. Lin, S. C. Liu, C. C.-K. Chou, T. H. Liu, C. T. Lee, C. S. Yuan, C. J. Shiu and C. Y. Young, Long-range transport of Asian dust and air pollutants to Taiwan, *TAO* **15** (2004) 759–784.
15. C. Y. Chan, L. Y. Chan, J. M. Harris, S. J. Oltmans, D. R. Blake, Y. Qin, Y. G. Zheng and X. D. Zheng, Characteristics of biomass burning emission sources, transport, and chemical speciation in enhanced springtime tropospheric ozone profile over Hong Kong, *J. Geophys. Res.* **108** (2003) doi:10.1029/2001JD001555.
16. M. O. Andreae, R. J. Charlson, F. Bruynseels, H. Storms, R. V. Grieken and W. Maenhaut, Internal mixture of sea salt, silicates, and excess sulfate in marine aerosols, *Science* **232** (1986) 1620–1623.
17. X. Li-Jones, H. B. Maring and J. M. Prospero, Effect of relative humidity on light scattering by mineral dust as measured in the marine boundary layer over the tropical Atlantic Ocean, *J. Geophys. Res.* **103**(D23) (1998) 31113–31121.

Advances in Geosciences
Vol. 10: Atmospheric Science (2007)
Eds. J. H. Oh and G. P. Singh

A HIGH-RESOLUTION SIMULATION OF CONVECTIVE-SCALE TRANSPORT OF DUST AEROSOL AND ITS REPRESENTATION IN CLOUD-RESOLVING SIMULATIONS

TETSUYA TAKEMI
Disaster Prevention Research Institute, Kyoto University, Gokasho, Uji, Kyoto 611-0011, Japan,
takemi@storm.dpri.kyoto-u.ac.jp

Boundary-layer processes under fair weather conditions play a significant role in the global budget of dust aerosols. Representing such small-scale processes are critical for better simulating dust transport in large-scale models. This study examines the microscale transport of dust aerosol due to convection by performing a high-resolution simulation and the sensitivity of the convective-scale dust transport to horizontal grid spacing of the order of 1 km typically used in cloud-resolving simulations. Turbulence effects play a significant role in raising and mixing dust aerosol within the boundary layer and cumulus convection transports dust further upward. A significant difference in the simulated dust transport is found between the simulations with the 4-km grid and the finer grid spacings. Through conducting the sensitivity simulations, we propose a simple formulation using an eddy velocity scale in order to represent both shallow and deep convection and hence to better reproduce dust emission and transport in cloud-resolving simulations.

1. Introduction

The transport of mineral dust aerosols in a desert region is regulated not only by synoptic-scale and mesoscale disturbances but also by smaller-scale processes such as boundary-layer turbulence and cloud-scale convection. Diurnal variability is significant for these small-scale processes especially under fair weather conditions. Luo *et al.*[1] showed that the diurnal variability of dust mobilization and concentration is responsible for 20% up to 80% of the total temporal variability over desert regions. Koch and Renno[2] estimated that boundary-layer convective motion contributes to about 35% of the global dust budget. The low-level transport due to convective injection from the planetary boundary layer (PBL) into the free troposphere across the Pacific accounts for a significant part in the total trans-Pacific transport.[3] Well-developed PBL and cumulus clouds, therefore, play an important role in controlling the transport of aerosols upward from the surface to the free troposphere.[4]

Such small-scale processes are critical from the viewpoint of simulating global-scale dust transport, since they are subgrid-scale fluctuations in a general circulation model (GCM) and hence their parameterizations are critical for the better representation of dust transport in GCMs.[5] Takemi *et al.*[6,7] investigated the role of boundary-layer and cumulus convection on the emission and transport of dust aerosols and showed that convection contributes to enhancing surface winds through downward momentum transport from the free troposphere. Considering that surface dust emission critically depends on the accuracy of surface wind prediction, a better representation of dry and moist convection is necessary for successful simulations.

Furthermore, in considering the transport of dust aerosol specifically over east Asian deserts, we should keep in mind that the PBL depth reaches up to 4 km,[8] and the ground elevation is 1–2 km above the mean sea level. Thus, dust aerosols may easily be entrained into the upper-level westerly jets to be transported in a long range. Therefore, an explicit representation of microscale processes is important in simulating the dust transport in the sense that the PBL processes contribute to enhancing the vertical transport.

In the present study, we perform a high-resolution simulation of dust aerosol transport due to shallow PBL convection and deep cumulus convection and investigate the sensitivity of the dust transport to grid resolutions by using the modeling framework of Takemi *et al.*[7] for idealized simulations of fair weather boundary layer. A horizontal grid spacing of 100 m is employed for the high-resolution simulation that explicitly resolves eddy motions in the PBL. Grid spacings of the order of 1 km, typically used in a cloud-resolving model (CRM),[9] are examined in the sensitivity experiments, because CRM simulations do not require cumulus parameterizations and will be employed in future regional-scale and global-scale simulations. Another purpose of this study is therefore to propose a simple representation of convective dust transport in CRM simulations through examining the resolution dependence. The case examined is a desert boundary layer in China, which Yasui *et al.*[10] observationally studied.

This chapter is organized as follows: In Sec. 2, model and experimental settings are described. In Sec. 3, the results of the 100 m simulation are at first presented as a reference for the sensitivity experiments that will be described next in this section. From the comparison between the reference run and the sensitivity experiments, a simple representation of convective-scale transport in cloud-resolving simulations is proposed in the last part of the section. The conclusions of this study are summarized in Sec. 4.

2. Model Setup and Experimental Design

The dust transport model used here is based on a regional meteorological model, the Advanced Regional Prediction System (ARPS),[11] which was developed by The University of Oklahoma Center for Analysis and Prediction of Storms. The emission and transport processes of dust aerosol have been built into the ARPS model.[12] The model is configured in an idealized way as in the study of Takemi *et al.*[6,7] in order to focus on the fundamental dynamics of convective dust transport under a fair weather condition. The meteorological model includes full physics parameterizations, i.e. cloud microphysics,[13] subgrid-scale (SGS) turbulence mixing,[14] land-surface physics, and radiative transfer.[15] In the dust module, the vertical dust flux at the surface is determined as the fourth power of friction velocity,[16] and the atmospheric transport is computed by the velocities obtained from the atmospheric model. The threshold of the friction velocity for dust emission is set to be $60\,\mathrm{cm\,s^{-1}}$. The dust property is represented as a mixing ratio, and a single size bin of 1.0-μm radius is assumed.

In order to perform high-resolution simulations that would explicitly resolve boundary-layer eddies, a large-eddy simulation (LES) model of Deardorff[14] is used for the parameterization of SGS turbulence mixing. The turbulence length-scale depends on the stability, and has the same value for both the horizontal and the vertical directions.

A high-resolution simulation with the horizontal grid spacing (Δx) of 100 m and the vertical grid spacings of 20–240 m (85 levels) is conducted in a mesoscale domain of 40 km (east–west, the x-axis) $\times$ 10 km (north–south, the y-axis) $\times$ 11 km (vertical, the z-axis). Although the spacing of 100 m seems to be relatively larger for LES, recent studies using this grid size have been successful in representing both shallow and deep convection;[17,18] therefore, we consider that the 100 m grid is sufficient for simulating both shallow and deep convection and the associated dust transport. A periodic condition is imposed at all the lateral boundaries, and the upper boundary is a rigid lid with a Rayleigh-type damping layer above the 9 km height. This 100 m grid simulation is referred to as the control.

In addition to the 100 m grid simulation, a series of experiments examining the sensitivity to horizontal grid spacing in the range of cloud-resolving simulations are performed. We set $\Delta x = 250$ m, 500 m, 1 km, 2 km, and 4 km. These simulations are conducted in a larger computational domain of 80 km × 20 km × 11 km, since a coarser-grid simulation will require a larger domain for resolving cloud-scales, i.e. a couple of kilometers. We have confirmed that the difference in computational area does not affect significantly the convection and transport processes by comparing the results with the two different computational areas in the case of 250 m grid. In addition, coarser vertical grid spacings of 20–650 m (36 levels) are used. In order to enhance the vertical mixing with these vertical grids, a type of nonlocal mixing scheme[19] that modify the turbulence length-scale in the vertical is employed.

The initial base state is set to the horizontal averages over the computational area after a three-day spin-up run with the 250-m grid starting with the vertical profile of Yinchuan, China, in the southern Gobi Desert, at 0600 LT (local time at this longitude) on 13 April 2002. After this spin-up run, a clear diurnal variation was represented. This base state, shown in Fig. 1, is used for initializing the model with random temperature perturbations added below the 1 km height, and has been used in our previous studies.[6,7] The model is integrated in time for 12 h for all the simulation cases. The analyses are conducted for the model outputs at 300-s interval.

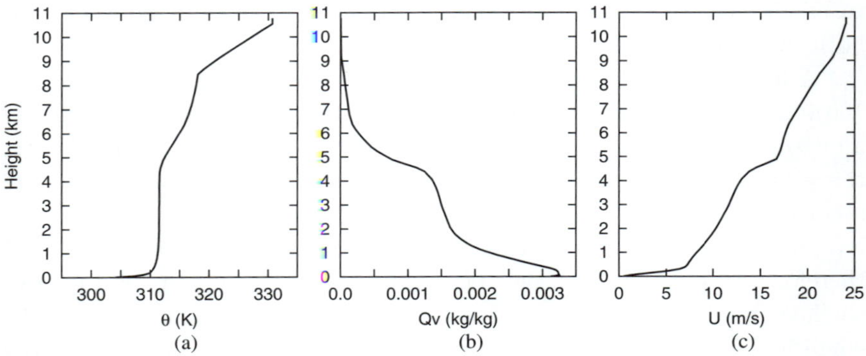

Fig. 1. Initial vertical profiles of (a) potential temperature (in K); (b) water vapor mixing ratio (in kg kg^{-1}); and (c) horizontal wind speed (in m s^{-1}, x-component).

3. Results

3.1. *High-resolution simulation*

The results of the control simulation are demonstrated as a reference for the sensitivity experiments.

Figure 2 shows the diurnal variation of the atmosphere in terms of stability and turbulence intensity. Virtual potential temperature and turbulent kinetic energy (TKE) are horizontally averaged over a computational area. The initial atmosphere has a strongly stable surface layer capped with a neutral layer (which is a residual of the previous daytime PBL); this initial stable layer is eroded by surface heating and disappears before 0900 LT. The turbulent intensity gradually increases from the surface, which leads to the formation of convective mixed layer. At 1000 LT, the top of the mixed layer reaches a height above 4 km and then deep convective clouds develop above the mixed layer. In the afternoon, the PBL convection remains to be strong, and deep convective clouds sporadically develop.

The diurnal variation of dust transport associated with the shallow and the deep convection is indicated in Fig. 3. Surface emission starts at 1000 LT when deep convective activity is intensified. The emitted dust is immediately raised within the PBL; the development of the dust layer agrees well with that of the layer with high TKE, which suggests that at

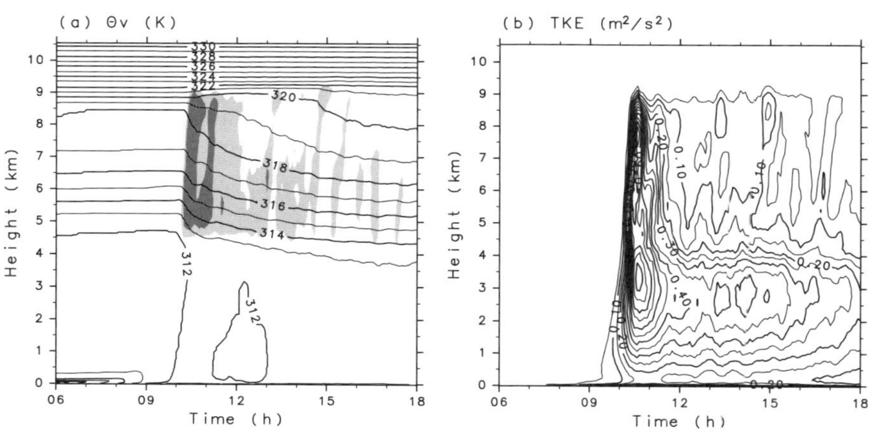

Fig. 2. Time–height section of horizontally averaged (a) virtual potential temperature (contoured at 1-K interval) and (b) turbulent kinetic energy (contoured at 0.05-$m^2\,s^{-2}$ interval). A cloud region defined as having cloud water plus ice mixing ratio of 0.025–0.05 $g\,kg^{-1}$ (lightly shaded) and greater than 0.05 $g\,kg^{-1}$ (darkly shaded) is indicated in (a).

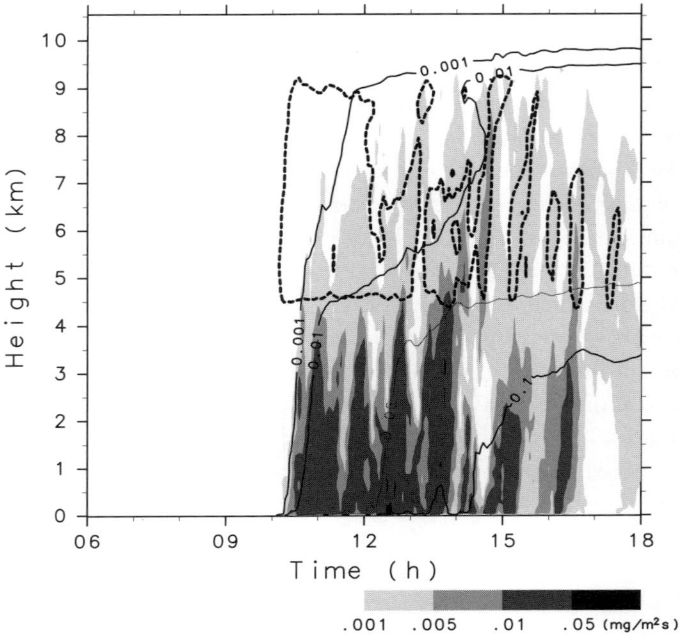

Fig. 3. Time–height section of area-averaged dust concentration (contoured at 0.001, 0.01, 0.05, and 0.1 mg m^{-3}) and vertical flux of dust (shaded, in mg m^{-2} s^{-1}). Cloud regions are depicted by dotted lines as a contour of cloud mixing ratio being 0.025 g kg^{-1}.

this stage the turbulent motion induces the transport of dust within the PBL depth. After 1100 LT the dust is further transported above the PBL owing to deep convective activity. In the afternoon, intense vertical flux of dust continues in the PBL, due to active turbulence associated with the PBL convection, and the dust content gradually increases. The transport into the free troposphere is also seen. Although the area-mean dust content is at most 0.2 mg m^{-3}, the maximum dust content during the simulation period amounts to 30.1 g m^{-3} at 1325 LT. This large difference between the mean and the maximum values suggests that a short-period variability due to turbulent and convective motion is critical in determining the peak values of atmospheric tracers.

3.2. *Resolution dependence*

In this subsection, the dependence of the simulated results to horizontal grid spacing is presented.

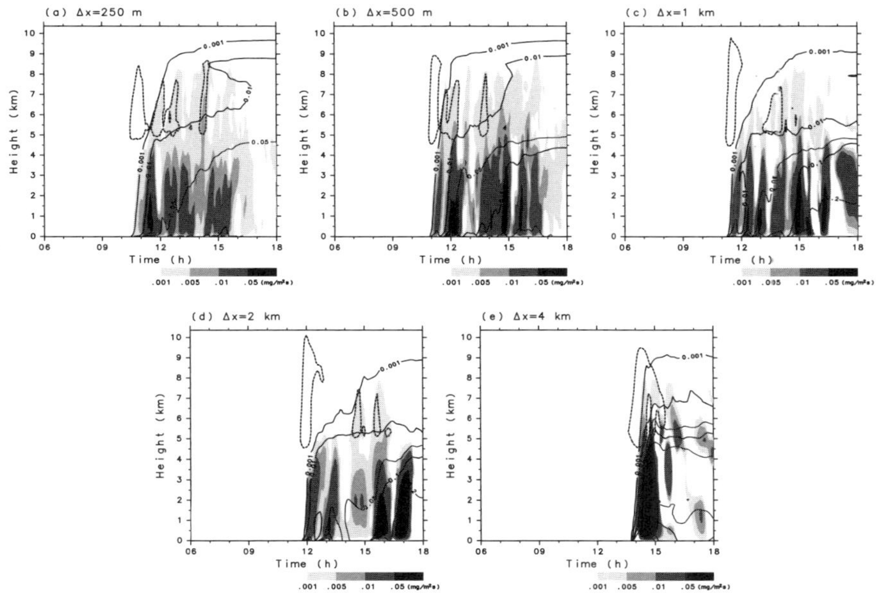

Fig. 4. The same as Fig. 3, except for the cases with Δx of (a) 250 m, (b) 500 m, (c) 1 km, (d) 2 km, and (e) 4 km.

Figure 4 shows the temporal and height variation of dust concentration, upward flux of dust, and cloud boundary for the cases with various grid spacings. The diurnal features found in the control simulation can also be seen for the cases with the coarser grids except the 4-km grid case in which a later but more sudden development of upward dust flux is seen. In the coarsest grid case, the unexpectedly sudden and intense cumulus convection during 1400–1500 LT resulted in the highest value of column integrated dust content among all the cases examined here. The column dust contents at the end of the simulation period for the control and the 500-m grid runs are $0.63\,\mathrm{g\,m^{-2}}$ and $0.80\,\mathrm{g\,m^{-2}}$, respectively, while the content for the 4-km run is $1.1\,\mathrm{g\,m^{-2}}$. It should be noted that the column content obtained by the control run is consistent with the estimation by satellite remote-sensing data.[20]

The resolution-dependence is further demonstrated by examining the difference of boundary-layer development. The PBL activity is diagnosed in terms of TKE that is computed in the SGS turbulence closure scheme.[14,19] Figure 5 compares the diurnal variation of the boundary-layer development for the cases with Δx = 250 m, 1 km, and 4 km. In the 250-m grid case, the PBL motion starts to intensify at 0900 LT, and the boundary layer

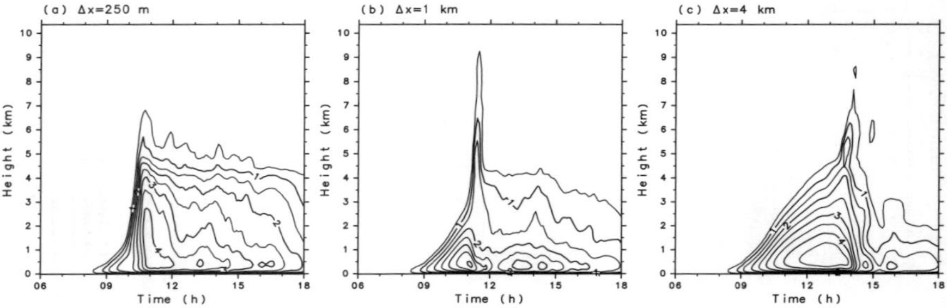

Fig. 5. The same as Fig. 4, except for turbulent kinetic energy (contoured at $0.5\,m^2\,s^{-2}$).

smoothly deepens, which leads to cumulus development at 1100 LT as seen in Fig. 4(a). A similar development can still be identified for the 1-km grid case. Note that the TKE values in Figs. 5(a) and 5(b) are significantly larger than those in the 100 m run (see Fig. 2(b)); this is due to the enhanced eddy viscosity, which takes into account the effects of nonlocal mixing in the sensitivity runs. On the other hand, although an increase of boundary-layer activity is seen for the 4-km grid case, the increase is much slower; thus, the cumulus development does not occur before noon and is significantly retarded. Considering that both boundary-layer and cumulus convection play a critical role in enhancing dust emission and transport,[6] the slow boundary-layer development in the coarsest grid case seems to be a reason for the significant difference in the dust transport from the finer-grid cases.

These results indicate that the grid spacing of 4 km, which is well in the range of explicit representation of deep convection in mesoscale systems,[21] is not sufficient to resolve shallow and deep convection under the present fair weather condition. In particular, the scales of updrafts for shallow convection are typically smaller than 4 km. The 4 km grid spacing is actually arguable, because there is no robust, satisfactory solution for convection and turbulence parameterization in the simulations.[22] The 4-km mesh, on the other hand, is obviously not coarse for regional-scale (let alone global-scale) simulations of convective dust transport, and hence most studies on atmospheric transport aim at performing simulations with grid spacings of 1–10 km; therefore, a proper parameterization for activating convection that induces dust emission and transport is necessary in this range of grid spacing.

One possibility for the better representation of the processes is to include a cumulus parameterization in the 4-km simulation. We further performed a 4-km simulation that included either the Kuo or the Kain–Fritsch cumulus parameterization built in the ARPS model,[15] and found that the boundary-layer development was not improved. This result is reasonable, since those parameterizations are not suitable for parameterizing both shallow and deep convections. Another possibility is to improve the TKE prediction by enhancing the production terms in the TKE closure equation; however, this requires a detailed analysis of a turbulence simulation data set. In the following, we examine a simple representation for facilitating the development of both boundary-layer and cumulus convection and hence enhancing dust production.

3.3. *Subgrid-scale updraft acceleration*

The TKE diagnosis in the previous subsection indicated that the diurnal evolution of turbulence intensity is slower in the 4-km simulation than in the finer-grid cases. In order to enhance the boundary-layer development in the 4-km case, therefore, the subgrid-scale turbulent effects need to be intensified, which will contribute to the acceleration of convective updrafts and downdrafts. Since turbulence intensity is increased especially in the updraft regions, updraft enhancement is considered to be critical in reproducing reasonable development of the PBL.

In order to develop a representation of convection effects in cloud-resolving simulations, we apply the idea of Deng *et al.*[23] who parameterize shallow convection in mesoscale models. They hypothesized that cloud-forming rising parcels were positively correlated with vertical velocity perturbations and proposed that an eddy vertical velocity should be added to resolved vertical motion that is related to activating cloud formation in a cloud microphysics parameterization; note that this enhanced updraft speed is not used anywhere except in this cloud activation computation. Such an adjustment to updraft velocity was also considered by Lohmann *et al.*[24] in activating cloud droplet nucleation. A point stressed here is that incorporating the turbulent effects into updraft speeds is critical in obtaining better results in simulations with Δx of a couple of kilometers.

Motivated by these studies, we assume that updraft is enhanced if there is a significant amount of SGS turbulent intensity. This assumption for updraft enhancement is based on the fact that updrafts are generally stronger than downdrafts in a single cell of shallow and deep convective clouds in which precipitation is not strong enough to produce cold air and intense downdraft. The enhancement of updrafts is conducted by adding a forcing term to the vertical momentum equation at updraft grid points. This term is referred to as subgrid-scale updraft acceleration (SUA), and is given by

$$\left(\frac{dw}{dt}\right)_{\mathrm{SUA}} = \frac{1}{\tau}\sqrt{\frac{2}{3}e}, \tag{1}$$

where w is vertical velocity, e is TKE, and τ is the timescale for the updraft acceleration. By trial and error we set $\tau = 10\,\mathrm{min}$. Although this forcing term is solely artificial, it is at least useful to examine the effects of turbulent intensity and hence to give an idea for the parameterization of convection effects in dust transport simulations.

Figure 6 shows the time–height section of TKE in the case of $\Delta x = 4\,\mathrm{km}$ with SUA. The result significantly improves the boundary-layer development looks quite similar to that seen in the finer-grid cases (compare with Figs. 2(b) and 5).

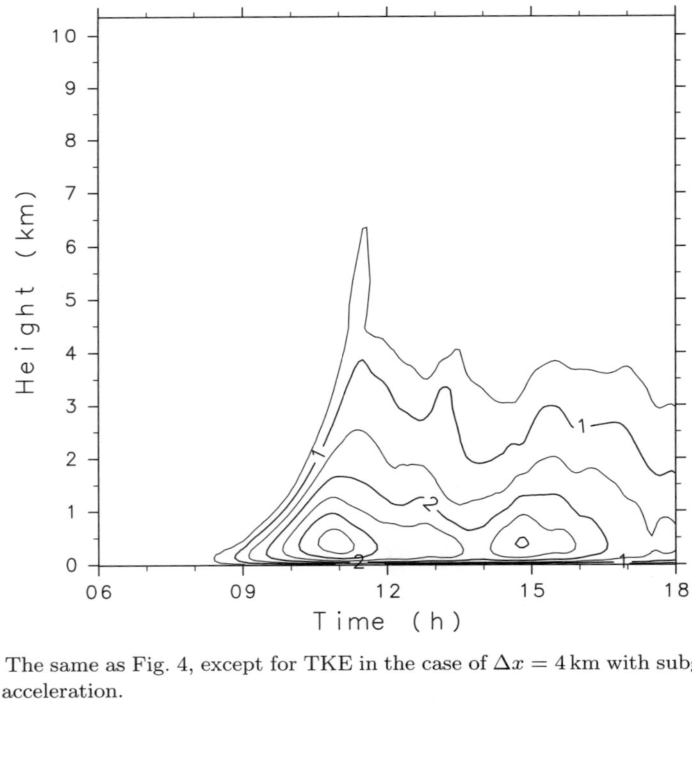

Fig. 6. The same as Fig. 4, except for TKE in the case of $\Delta x = 4\,\mathrm{km}$ with subgrid-scale updraft acceleration.

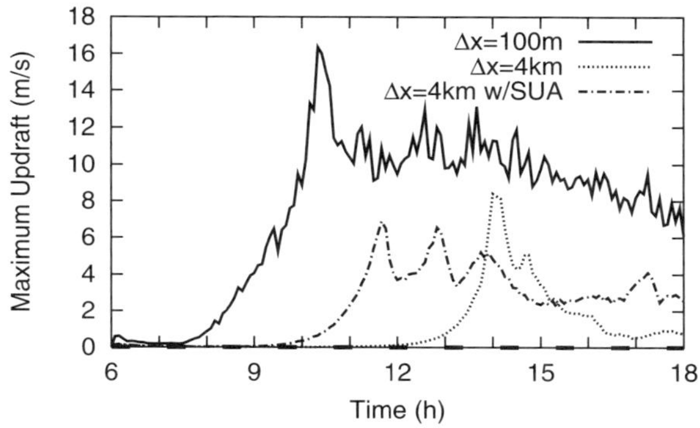

Fig. 7. Temporal variation of the maximum updraft in the domain for the cases of 100 m grid, 4-km grid, and 4-km grid with subgrid-scale updraft acceleration.

The temporal variation of the maximum upward velocity in the computational domain is compared in Fig. 7. Although the velocity in the 4-km grid run with SUA is weaker than that in the 100-m grid case, adding the forcing term overall improves the velocity variation as compared with the 4-km run without the parameterization. Furthermore, the distribution of the frequency of surface wind speed (not shown) indicated that the feature seen in the 100-m grid case was better captured in the case of 4-km grid with SUA than without the parameterization.

With the boundary-layer development and associated surface wind variation being better represented with the SUA parameterization, the convective dust transport therefore is better reproduced as shown in Fig. 8 (compare with Fig. 4(e)). The diurnal evolution of vertical dust transport with SUA compares well with the control case as shown in Fig. 3: the initial development and gradual increase of dust content is better represented with the parameterization than without it. The column dust content at the end of the simulation period is $0.72\,\mathrm{g\,m^{-2}}$; this value also compares well with the estimate by the control simulation.

In this way, the simple formulation of Eq. (1) seems to be a plausible candidate for parameterizing both shallow and deep convection under the present meteorological setting and for representing the associated convective dust transport. However, a proper choice of τ remains to be determined. We consider that τ should be determined from a characteristic timescale for convective motion with a scale of O (1 km). A typical velocity for this motion is on the order of $1\,\mathrm{m\,s^{-1}}$, and thus the equivalent timescale becomes $\sim$1000 s. Therefore, the present choice of $\tau = 10\,\mathrm{min}$

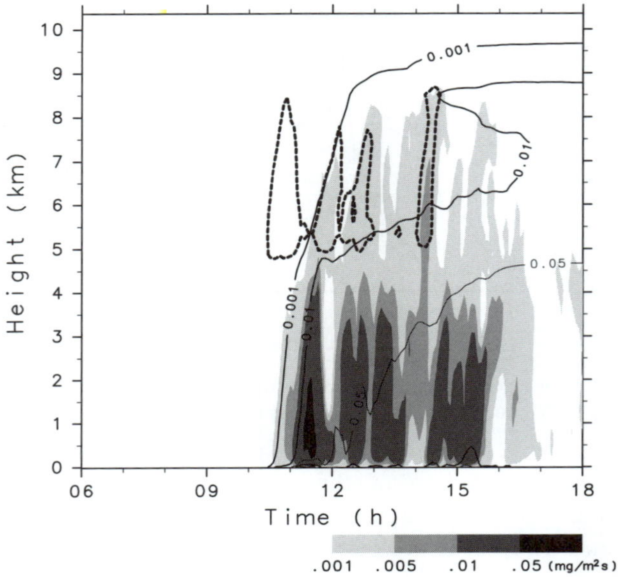

Fig. 8. The same as Fig. 3, except for the case of $\Delta x = 4\,\mathrm{km}$ with subgrid-scale updraft acceleration.

is considered to be reasonable. Overall, introducing the parameterization seems to improve the development and evolution of dust transport.

4. Summary and Conclusions

This study investigated the microscale transport of dust aerosol due to convection under a fair weather condition in a desert area. A high-resolution (i.e. $\Delta x = 100\,\mathrm{m}$) simulation that explicitly resolves shallow and deep convective motions has been conducted. A diurnal variation of PBL and cumulus development were well reproduced. Dust aerosol is raised and mixed within the PBL by intense turbulent motion, and is transported further upward in the free troposphere by deep cumulus convection.

We have also performed a series of numerical experiments examining the sensitivity of convective dust transport to model resolution in the cloud-resolving range with a grid spacing of $O\,(1\,\mathrm{km})$. The case examined was a fair weather boundary layer and an associated cumulus development over a desert area. Since boundary-layer organized convection has a spatial scale of $O\,(1\,\mathrm{km})$,[10] the 4 km grid was not sufficient to resolve the convective motion, and a significant gap was found in the representation of both shallow and deep convection between the cases with the 4-km grid and with the finer grids. The simulated results suggest that a proper parameterization for activating convection is required for the 4-km simulation.

By examining the resolution-dependence, we presented a simple formulation using TKE for enhancing updraft and demonstrated that the coarse-mesh simulation with this formulation significantly improves the convective dust transport. Since the present modeling framework is configured in an idealized way with full physics processes included as in Takemi *et al.*,[6,7] other complicating factors such as the data assimilation technique and lateral boundary condition need not be taken into account.

Considering that a simulation with a grid size of 1–10 km in the cloud-resolving range will be affordable at least in the regional-scale, a parameterization for activating both shallow and deep convection should be included in the CRM simulations. The most important point to be stressed in this study is that the use of an eddy velocity scale, in the present formulation or the other, is plausible for better representing the cumulus-scale transport of dust aerosol. The present results further suggest that the production of TKE in the closure scheme should be enhanced in the 4-km simulation. A future study should conduct high resolution simulations of both shallow and deep convection under various meteorological settings in order to examine the validity of the present formulation and to revise the parameterization based more on physics processes in the boundary layer.

Acknowledgments

This work was supported partly by Grant-in-Aid for Scientific Research 19740287 from the Japan Society for the Promotion of Science. The GFD Dennou Library (http://www.gfd-dennou.org/index.html.en) was used for drawing some of the figures.

References

1. C. Luo, N. Mahowald and C. Jones, Temporal variability of dust mobilization and concentration in source regions, *J. Geophys. Res.* **109** (2004) D20202, doi: 10.1029/2004JD004861.
2. J. Koch and N. O. Renno, The role of convective plumes and vortices on the global aerosol budget, *Geophys. Res. Lett.* **32** (2005) L18806, doi:10.1029/2005GL023420.

3. M. Holzer and T. M. Hall, Low-level transpacific transport, *J. Geophys. Res.* **112** (2007) D09103, doi:10.1029/2006JD007828.
4. J. K. S. Ching and A. J. Alkezweeny, Tracer study of vertical exchange by cumulus clouds, *J. Climate Appl. Meteorol.* **25** (1986) 1702–1711.
5. R. V. Cakmur, R. L. Miller and O. Torres, Incorporating the effect of small-scale circulations upon dust emission in an atmospheric general circulation model, *J. Geophys. Res.* **109** (2004) D07201, doi:10.1029/2003JD004067.
6. T. Takemi, M. Yasui, J. Zhou and L. Liu, Modeling study of diurnally varying convective boundary layer and dust transport over desert regions, *Sci. Online Lett. Atmos.* **1** (2005) 157–160.
7. T. Takemi, M. Yasui, J. Zhou and L. Liu, The role of boundary-layer and cumulus convection on dust emission and transport over a midlatitude desert area, *J. Geophys. Res.* **111** (2006) D11203, doi:10.1029/2005JD006666.
8. T. Takemi and T. Satomura, Numerical experiments on the mechanisms for the development and maintenance of long-lived squall lines in dry environments, *J. Atmos. Sci.* **57** (2000) 1718–1740.
9. W. W. Grabowski, Coupling cloud processes with the large-scale dynamics using the cloud-resolving convection parameterization (CRCP), *J. Atmos. Sci.* **58** (2001) 978–997.
10. M. Yasui, J. Zhou, L. Liu, T. Itabe, K. Mizutani and T. Aoki, Vertical profiles of aeolian dust in the desert atmosphere observed by a lidar in Shapotou, China, *J. Meteorol. Soc. Jpn.* **83A** (2005) 149–171.
11. M. Xue, K. K. Droegemeier and V. Wong, The Advanced Regional Prediction System (ARPS) — A multiscale nonhydrostatic atmospheric simulation and prediction tool. Part I: Model dynamics and verification, *Meteorol. Atmos. Phys.* **75** (2000) 161–193.
12. T. Takemi, Explicit simulations of convective-scale transport of mineral dust in severe convective weather, *J. Meteorol. Soc. Jpn.* **83A** (2005) 187–203.
13. Y. L. Lin, R. D. Farley and H. D. Orville, Bulk parameterization of the snow field in a cloud model, *J. Climate Appl. Meteorol.* **22** (1983) 1065–1092.
14. J. W. Deardorff, Stratocumulus-capped mixed layers derived from a three-dimensional model, *Bound. Layer Meteorol.* **18** (1980) 495–527.
15. M. Xue, K. K. Droegemeier, V. Wong, A. Shapiro, K. Brewster, F. Carr, D. Weber, Y. Liu and D. Wang, The Advanced Regional Prediction System (ARPS) — A multiscale nonhydrostatic atmospheric simulation and prediction tool. Part II: Model physics and applications, *Meteorol. Atmos. Phys.* **76** (2001) 143–165.
16. M. Liu and D. L. Westphal, A study of the sensitivity of simulated mineral dust production to model resolution, *J. Geophys. Res.* **106** (2001) 18099–18112.
17. M. Khairoutdinov and D. Randall, High-resolution simulation of shallow-to-deep convection transition over land, *J. Atmos. Sci.* **63** (2006) 3421–3436.
18. A. Cheng and K.-M. Xu, Simulation of shallow cumuli and their transition to deep convective clouds by cloud-resolving models with different third-order turbulence closures, *Quart. J. Roy. Meteorol. Soc.* **132** (2006) 359–382.

19. W. Y. Sun and C. Z. Chang, Diffusion model for a convective layer. Part I: Numerical simulation of convective boundary layer, *J. Climate Appl. Meteorol.* **25** (1986) 1445–1453.
20. M. Kuji, N. Yamanaka, S. Hayashida, M. Yasui, A. Uchiyama, A. Yamazaki and T. Aoki, Retrieval of Asian dust amount over land using ADEOS-II/GLI near UV data, *Sci. Online Lett. Atmos.* **1** (2005) 33–36.
21. M. L. Weisman, W. C. Skamarock and J. B. Klemp, The resolution dependence of explicitly modeled convective systems, *Monthly Weather Rev.* **125** (1997) 527–548.
22. A. Deng and D. R. Stauffer, On improving 4-km mesoscale model simulations, *J. Atmos. Sci.* **60** (2006) 34–56.
23. A. Deng, N. L. Seaman and J. S. Kain, A shallow-convection parameterization for mesoscale models. Part I: Submodel description and preliminary applications, *J. Atmos. Sci.* **60** (2003) 34–56.
24. U. Lohmann, N. McFarlane, L. Levkov, K. Abdella and F. Albers, Comparing different cloud schemes of a single column model by using mesoscale forcing and nudging technique, *J. Climate* **12** (1999) 438–461.

Advances in Geosciences
Vol. 10: Atmospheric Science (2007)
Eds. J. H. Oh and G. P. Singh

ICE-NUCLEATING AND OPTICAL PROPERTIES OF ICE CLOUD SEEDED BY DIMETHYL SULFOXIDE (DMSO)

L. N. BISWAS*, A. HAZRA†, P. MAITI*, V. MANDAL*,
U. K. DE*,§ and K. GOSWAMI‡

**Atmospheric Science Research, Environmental Science Programme, Jadavpur University, Kolkata 700032, India*

†Department of Atmospheric Science, National Taiwan University, Taipei, Taiwan 106, ROC

‡Department of Physics, Jadavpur University, Kolkata 700032, India
§deutpal2003@yahoo.com

The widespread presence of DMS (Dimethyl sulfide) in sea water and its atmosphere is now established. DMS may be oxidized into DMSO (Dimethyl Sulfoxide), and its presence in the atmosphere is also established.

DMSO was seeded inside a cold room to examine its ice-nucleating characteristics and to study the optical properties of the iced cloud. It was found to be a good ice-nucleating agent. The maximum nucleation occurred at −21.0°C, which is the eutectic temperature of DMSO and water bivariate system. The variation of modal value of the dimension of crystal with seeding temperature had consistent pattern with the count of crystal nucleation. When DMSO-coated Ammonium Sulfate dust was seeded, two peaks in nucleation count appeared. While the highest peak was at −21.0°C, like the previous experiment, the second highest peak occurred at −18.2°C, which is the eutectic temperature of Ammonium sulfate and water combine. However, the crystal count increases manifold. In this case also, the modal value of crystal size had a consistent pattern with the variation of crystal count.

To study the optical properties, a laser beam of wavelength 633 nm is sent through the iced cloud. In general, the forward-scattering intensity is greater than the backward-scattering intensity by an order of 1 or 2. But, from −18°C to −13°C, the backward-scattering dominates over the forward-scattering. This feature is attributed to halo effect, as the cubic and hexagonal shaped ice crystals dominate then. Apart from that, the variation of scattering coefficient, extinction coefficient, and optical thickness of the ice cloud, with seeding temperature is also presented.

§Corresponding author.

1. Introduction

The element sulfur is essential for the sustenance of life. So it is necessary to take stock of supply of sulfur in the sulfur cycle. There are a number of natural sources of atmospheric sulfur. Apart from volcanic emission and the Aeolian generation of sulfate particles, the biogenic emissions from land and sea are important sources.[1] At one time, H_2S was considered as the main representative of biogenic source, as it is copiously produced in anaerobic marshlands and tidal flats. However, since the early 1970s, the scenario has changed. A number of additional sulfides were discovered in the atmosphere in that decade, like carbonyl sulfide (OCS),[2] carbonyl disulfide (CS_2),[3] dimethyl sulfide (DMS) (CH_3SCH_3)[4] and dimethyl disulfide ($CH_3S_2CH_3$).[5] In 1972 Lovelock *et al.*[6] first discovered that sea water is saturated with DMS and then, in 1977, Maroulis and Bandy[4] discovered it in marine atmosphere. After the works of Nguyen *et al.*[7] and Barnard *et al.*,[8] the ubiquitous presence of DMS in the atmosphere was established.

In the atmosphere, DMS is supposed to be oxidized into SO_2 and methane sulfonate. SO_2 is converted into sulfuric acid and particulate sulfate. However, laboratory experiments identify that there may be various other important end products during oxidation. These are HCHO, methane sulfonic acid (MSA: CH_3SO_3H), dimethylsulfoxide (DMSO: CH_3SOCH_3), and methane sulfinic acid (MSIA: $CH_3S(O)(OH)$).[9,10] All sulfides except OCS react rapidly with OH radicals. DMS reacts with OH in two pathways, either by abstraction of a hydrogen atom or by addition of OH to sulfur atom (Ref. 1, p. 603):

$$\begin{aligned} OH + CH_3SCH_3 &\rightarrow H_2O + CH_3SCH_2 \\ &\leftrightarrow CH_3S(OH)CH_3 \end{aligned}$$

In the presence of O_2, the addition product reacts with it to produce DMSO[11]:

$$CH_3S(OH)CH_3 + O_2 \rightarrow CH_3SOCH_3 + HO_2$$

DMSO has also been produced in the laboratory by photochemical reactions of DMS in the presence of OH.[12] There exist reports about the possibility of formation of DMSO due to microbial oxidation and photo-oxidation of DMS, as well as due to the direct biosynthesis by phytoplanktons.[13]

In fact, both DMS and DMSO have been detected in the marine atmosphere.[14] Many sulfates or sulfur-containing aerosols are now recognized as ice or cloud condensation nuclei (IN or CCN).[15–18] In the present work, DMSO is identified as a good IN.

The behavior of DMSO as IN has been studied inside a laboratory scale cold room over the negative side of 0°C starting from −22°C on the lower end. The ice nucleation ability is found to be maximum at −21.0°C, which is also the eutectic temperature of DMSO and water bivariate system. In fact, this observation is consistent with what has been noted in the case of many other seeding agents.[18–21] Above that temperature, the nucleation ability gradually falls off and almost vanishes near 0°C. The mode of the dimension of falling crystals at different seeding temperatures was noted, and the modal value has consistent temperature-variational pattern with the crystal count. No nucleation ability is noted above 0°C, unlike the agents like ammonium sulfate,[18] benzoin dust,[22] urea,[20] or the bacteria *Pseudomonas aeruginosa*.[23]

In reality some of the DMSO particulates are supposed to be carried to the middle and upper atmosphere by convection, and these have the ability to form heterogeneously nucleated ice cloud. The concerned clouds can scatter and absorb solar radiation as well as the radiation emitted from the earth, and have the ability to modify the solar radiation budget of the earth. In order to estimate the extent of modification to the radiation budget due to DMSO, study of bulk-scattering properties and absorption characteristics of ice crystals formed at different temperatures and over the entire spectrum of solar radiation is necessary. This particular exercise is partially taken up in the present study.

DMSO exists dominantly near the sea surface, and it can exist as liquid drop particularly over the tropical sea region. The liquid drops of DMSO may coat many solid aerosols. As a large fraction of the aerosols are in the form of sulfates,[24] the effect of coating of a dominant sulfate i.e. Ammonium Sulfate, by DMSO has also been studied.

To study the optical properties of ice cloud nucleated by DMSO, a 5 mW plane-polarized He–Ne laser beam (wavelength 633 nm) was passed through it. The observational procedures were almost identical to a similar work done for sodium chloride dust and aqueous solution.[21] The time for crystal growth and fall out can be estimated from the rise and fall of scattering intensity at different angles with time. The variation of scattering intensity at different angles with change of seeding temperature is noted. The scattering coefficient, extinction coefficient, and optical depth are also evaluated.

2. Experiment

The experiment was carried out inside a single-door cold room of dimensions 1.52 m, 1.52 m, and 1.83 m as length, breadth, and height, respectively (Fig. 1). The body of the room is made of galvanized iron sheet and is insulated by PUF. The temperature of the room can be reduced up to −35°C. An air-conditioner is operated outside the room so that the temperature difference between outside and inside the cold room can be reduced. One thermistor is used to note the temperature inside the cold room and another is used to note that of the outside. Both temperatures have digital display. Inside the room, a spherical glass vessel of diameter 35.6×10^{-3} m and 20×10^{-3} m^3 capacity is placed, and it has several outlets on its body.

A cloud of supercooled droplets was produced by cooling the vessel and passing steam through a hole near the floor of the vessel. A closed container with water was used to produce steam and it entered the vessel through a pipeline. The amount of water vapor inside the vessel was controlled by adjusting the steam flow. The seeding material was injected inside the vessel through a port. Two thermistors were used to measure the temperature inside the vessel. Of the two, one was kept near the floor of the vessel and the other, kept just above the seeding port. The seeding temperature being mentioned in the literature was that noted in the second thermistor, as nucleation mostly occurs in its neighborhood. Two thermistors were used to find the temperature gradient inside the vessel. All the thermistors were

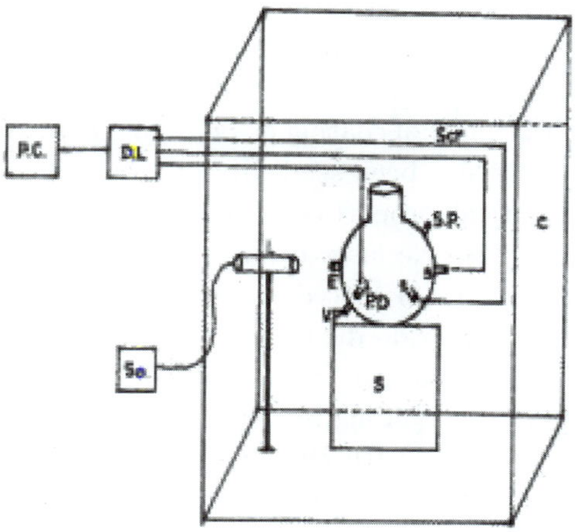

Fig. 1. Schematic arrangement of experimental apparatus inside the cold room.

calibrated at ice and steam points, and the calibration error was within 0.1°C. The temperature difference between the two sensors within the vessel remained within 0.1°C, when a steady condition was reached.

One circulator was fixed at the roof of the cold room to keep the temperature uniform inside the room. A light was also kept inside the room for working purposes.

At first, the temperature of the room was brought to a pre-decided value of low temperature, and then steam was passed into the vessel until a steady temperature was reached. This steady condition can be controlled in two ways, i.e. by fixing the temperature of the vessel when steam was passed and controlling the flow of steam. The steady temperature can be retained for about 5 min. The seeding was done at that steady temperature. The entire experiment was performed within this steady condition. One should note that, as steam was passed, the temperature of the room remained almost constant, but that of the vessel underwent quick change. In all the experiments the total water content of the vessel was kept fixed within a narrow range of $130-150 \times 10^{-3}$ kg.

Commercially available DMSO being used here is 99.5% chemically pure. As the freezing point of DMSO is 17–18°C, it remains in liquid state at the normal temperature of the laboratory. A volume of 20×10^{-6} m^3 of the liquid, brought quickly from room temperature, was sprayed inside the vessel in one stroke. The syringe being used here had a nozzle and admitted only droplets having diameter with a peak of 0.7 μm and standard deviation of 0.21 μm. It is expected that DMSO will form solid particulate inside the chilled vessel.

To study the optical property, the laser beam was passed through a port located in the middle of the vessel. Five other ports are located in the same horizontal section of the vessel so that the directly scattered beam as well as those scattered at angles 30°, 36°, 144°, and 150° with respect to the forward direction can be received. Five photodiodes are placed behind the five ports to receive the scattered light. A current-regulated circuit then amplifies the signal, and the amplified voltage output then goes to a data logger. Finally, all the data registered in the data logger are transferred to a PC for archiving. Immediately after the seeding the scattered intensities are found to change with time. The scattered intensity is automatically noted and archived at an interval of 1 s. When the voltage output returns to the base value, one can assume that the medium has become clear of ice cloud.

Crystals formed within the ice cloud finally settle down at the base of the vessel. To collect the falling-crystals formvar-coated glass slides were used. Formvar is a brand name of polyvinyl formal resin. A solution was prepared with 0.4×10^{-3} kg of formvar in $10 \times 10^{-6}\,m^3$ of chloroform. A thin layer of this solution was coated in two slides at a time, and these were kept in a tray near the floor of the vessel to collect the signature of the crystals falling there. The lid over the tray can be opened or closed using external control. Crystals were collected for 2 min only after seeding was done. When the slides were brought outside after the experiment, the chloroform quickly evaporated leaving behind a plastic film of formvar. On the other hand, the ice part of crystals melted to water. Details about these formvar-coated slides are already available in the literature.[25,26]

The slides were then kept inside a desiccator with sufficient silica gel, so that water would evaporate and be quickly absorbed by the gel. A replica of the arrested crystals is then retained in the slide, and in the case of heterogeneous nucleation the nucleating agent is left behind at the core. One can observe the replicas through a microscope.

The concentration, size, and basic habits of the crystals in a slide were noted at different temperatures. The microscope being used here has a magnification of 400. On each slide, 10 randomly chosen fields of view are taken for counting.

To study the nucleation ability of DMSO-coated Ammonium Sulfate dust, 5 g of Ammonium Sulfate dust was coated with 0.7 g of DMSO in liquid state. Ammonium Sulfate dust had dimension in the range of 0.1–0.7 μm. In fact, the above-mentioned amount of DMSO was just sufficient to coat fully the dust particulates of Ammonium sulfate. An amount of 20×10^{-6} kg of coated Ammonium Sulfate was sprayed inside the vessel in one stroke, keeping consistency with the first experiment. All the conditions were kept identical like the experiment with DMSO only. Like the first experiment, the concentration, size, and basic habits of the crystals formed in a slide were noted at different temperatures.

3. Theory for Optical Parameters

A detailed theory for the optical parameters is already available in the literature.[21] The effective volume scattering coefficient, when the angle of scattering is θ, is evaluated following the procedure given there.

The extinction coefficient is determined using Beer's law,

$$I = I_0 e^{-\sigma T},$$

where I = Intensity of the direct beam after scattering,
I_0 = Initial intensity of the direct beam,
σ = extinction coefficient,
T = thickness (here, it is the diameter of the vessel).
So, the extinction coefficient, $\sigma = (1/T)\ln(I_0/I)$, and the optical thickness is $\tau = \sigma T = \ln(I_0/I)$.

4. Experimental Results and Discussion

4.1. *Variation of crystal count and dimension with seeding temperature in case of DMSO*

The variation of crystal count with seeding temperature is represented in Fig. 2. At each value of seeding temperature, the seeding was done four times, the range of which was at best within 0.4°C. The seeding temperature presented in the figure was actually the mean of the four temperature values. There were two slides, and in each slide, 10 random observations were made by the microscope. The modal value of the dimension of all crystals measured in 80 fields of view of the microscope at a particular seeding temperature is also presented in the same figure. The variation

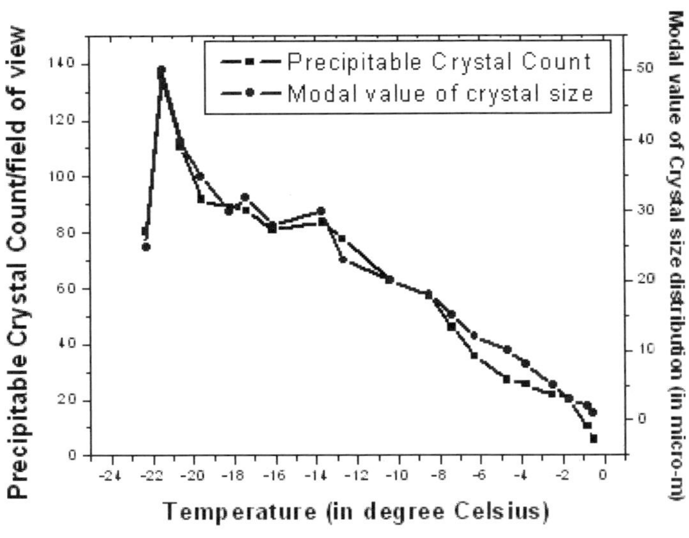

Fig. 2. Crystal count and modal value of crystal dimension against seeding temperature when DMSO is seeded.

of crystal count and the modal value of crystal dimension with seeding temperature have high degree of consistency. As already stated, the seeding temperature is varied from −22.0°C to 0°C. Both, the crystal count and the modal value of dimension have a sharp peak at −21°C, which is also the eutectic temperature of DMSO and water bivariate system. As Gibb's free energy becomes minimum at the eutectic point,[27] there is a common tendency for freezing at that temperature. The count as well as the modal dimension show a sharp rise as the temperature is raised from −22°C to −21°C. On the other hand, as the seeding temperature is raised above −21°C, both the crystal count and the modal dimension show a gradual fall. However, after a sharp fall in crystal count up to −19°C, the fall becomes more gradual. Except that, close to −14°C, there occurs a small peak in the crystal count, and at 0.5°C, both these quantities show insignificant values. From the nature of the curve, one can conclude that DMSO nuclei have no nucleation above 0°C.

On the other hand, in the case of mode of crystal dimension, a sharp fall in value occurs as the temperature rises from −21°C to −18°C. There are two small peaks in the modal value at −17°C and close to −14°C.

4.2. *Variation of crystal habit with seeding temperature in case of DMSO*

The overall variation of crystal habit with a change of seeding temperature is presented in Table 1. The observation is consistent with the fact that the principal factor governing the basic habit of ice crystal structure is temperature.[28] Rogers and You[29] have also concluded that the basic habits

Table 1. Crystal habits when DMSO is seeded.

Temperature	Crystal habit
−22°C to −17°C	Dominantly dendrite (Photo 1), some are hexagonal and cubic [Dendrite (80.54%): 9.12–45.56 μm; Hexagonal (10.16%): 4.25–25.12 μm; Cubic (9.3%): 5.13–11.24 μm]
−17°C to −15°C	Mostly cubic (Photo 2), some are hexagonal [Cubic (85.35%): 4.55–40.14 μm; Hexagonal (14.65%): 25.2–45.5 μm]
−15°C to −7°C	Mostly hexagonal (Photo 3), some are cubic or prism like [Hexagonal (88.12%): 25.32–50 μm; Cubic or prism (11.88%): 10.25–45.32 μm]
−7°C to −0.5°C	Almost all crystals are cubic or rod (Photo 4): (1.24–11.32 μm)

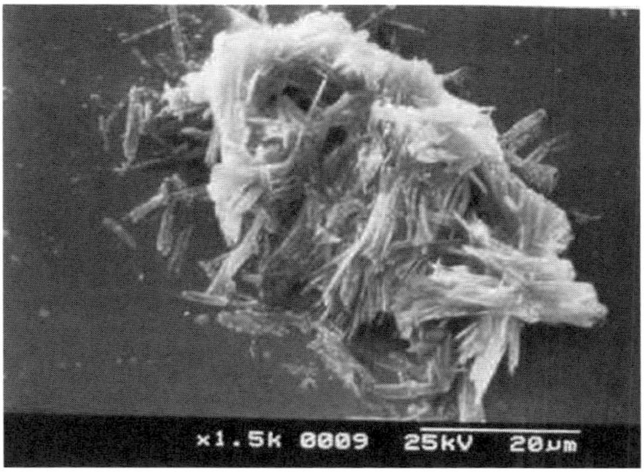

Photo 1. Dendrite formation at −18.0°C.

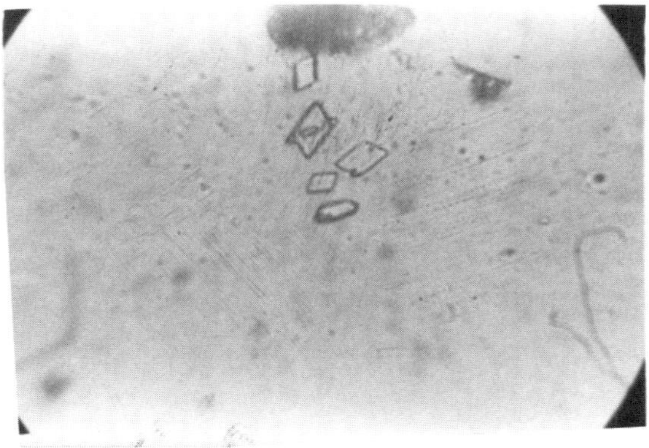

Photo 2. Cubic crystals at −16.0°C.

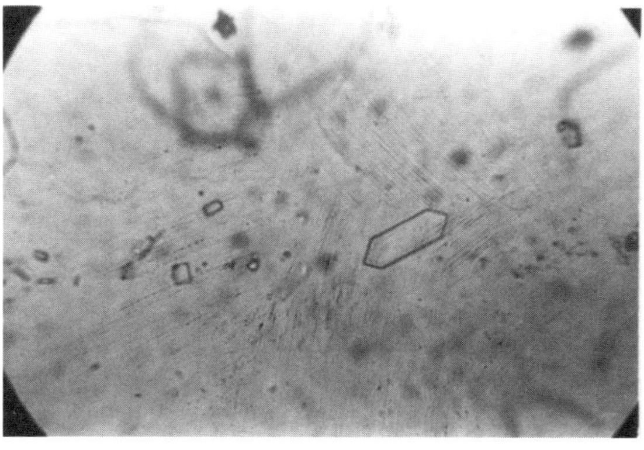

Photo 3. Single hexagonal crystal at −14.4°C.

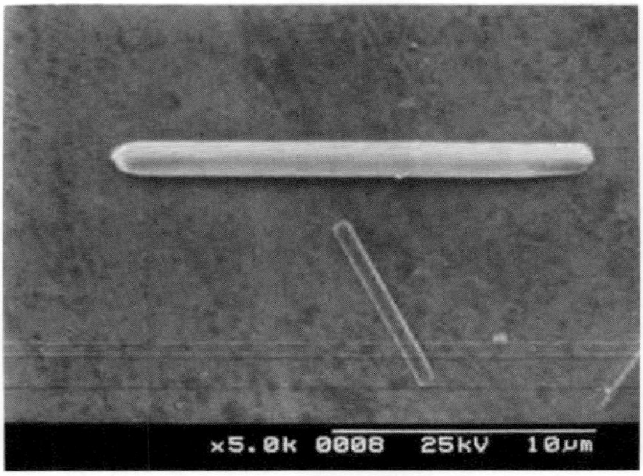

Photo 4. Large rod shaped crystal at −4.2°C.

of ice crystals are dependent on temperature, and they are independent of seeding materials.

4.3. *Identification of coating of Ammonium Sulfate dust with DMSO by NMR*

In order to confirm that the dust particles of Ammonium Sulfate are really coated with DMSO, the coated crystals were subjected to ^{1}H-NMR study in D_2O solvent. In the NMR spectrum (Fig. 3), we observe a sharp peak at

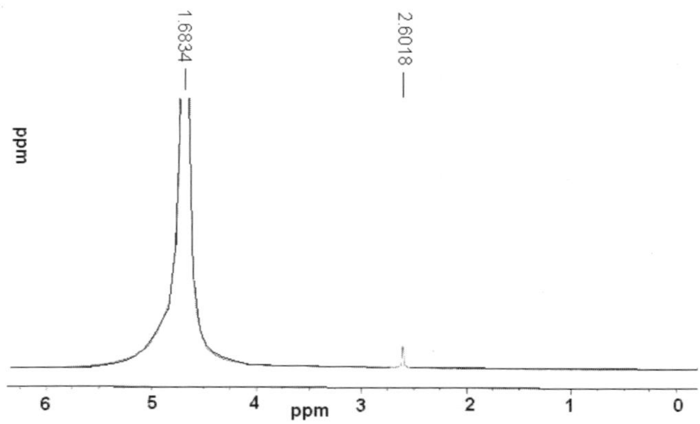

Fig. 3. NMR spectrum of DMSO-coated Ammonium Sulfate crystals.

δ 4.683, because of the presence of H_2O as an impurity along with D_2O. The other peak at δ 2.60 is due to DMSO. This clearly shows that Ammonium Sulfate dust particulates are actually coated with DMSO.

4.4. *Variation of crystal count and dimension with seeding temperature in case of DMSO-coated Ammonium Sulfate*

The variation of crystal count with seeding temperature for the second experiment is presented in Fig. 4. The modal value of dimension of all the crystals measured in 80 fields of view of the microscope at a seeding temperature is also presented in the same figure. The crystal count and modal value of dimension have significant consistency with the variation of seeding temperature. Like the previous experiment, the seeding temperature was varied from −22.0°C to 0°C. The most dominant distinction between the two experiments is in the count of crystals. The crystal count has gone up many folds particularly at the peak values, though the crystal dimension remained almost unaltered. The highest peak in the crystal count occurred at −21.0°C, which is the eutectic temperature of DMSO and water combine. The second highest peak in crystal count occurred at −18.2°C, which is the eutectic temperature of Ammonium

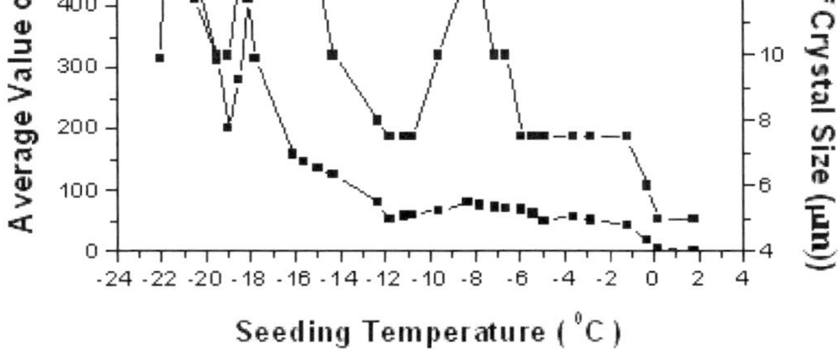

Fig. 4. Variation of crystal count and modal value of crystal dimension with seeding temperature in case of DMSO-coated Ammonium Sulfate.

Sulfate and water combine. However, the first crystal count was almost twice the second count. Beyond the temperature of the second peak toward 0°C, the count gradually goes down.

The modal dimension of crystal count has two distinct peaks coincident with the two peak temperatures in the crystal count. But, the two peaks in the modal value attain almost the same height. There is a distinct third peak in the modal value around −8°C; the reason of its existence is not at all clear.

4.5. *Variation of crystal habit with seeding temperature in case of DMSO coated with Ammonium Sulfate*

The variation of crystal habit with temperature is presented in Table 2. The variation of basic habit with temperature is consistent to a large extent with Table 1, except for the shift in percentage and size. In this case, dendrite formations have much smaller size and the rod-shaped formations exist over a wider range of temperatures.

4.6. *Optical parameters when DMSO is seeded*

In clear conditions the intensity of direct beam is always 13.4 J/m only. However, after the introduction of steam, the same intensity drops to a value about 6 J/m only. In case of seeding, the scattered beams are produced at the expense of direct intensity. The forward-scattered intensities at 30°

Table 2. Crystal habits when DMSO-coated Ammonium Sulfate is seeded.

Temperature	Crystal habit
−22°C to −17°C	Dendrite formations are highest. Rod and rectangular shapes are also significant. [Dendrite (43%): 12.5–17.5 μm; Rod (32%): 12.5–17.5 μm; Rectangular (25%): 10–15 μm]
−17°C to −8°C	Rectangular formations are dominant. Rod, dendrite, hexagonal, and cubic forms are also present. [Rectangular (61%): 5–7 μm; Rod (19%): 15–50 μm; Dendrite (10%): 8–12 μm; Hexagonal (6%): 5–12 μm; Cubic (4%): 7–15 μm]
−8°C to 0°C	Rod shapes are dominant. Cubic, rectangular, and hexagonal are other formations. [Rod (53%): 5–12 μm; Cubic (19%): 7–10 μm; Rectangular (18%): 5–12 μm; Hexagonal (10%): 10–20 μm]

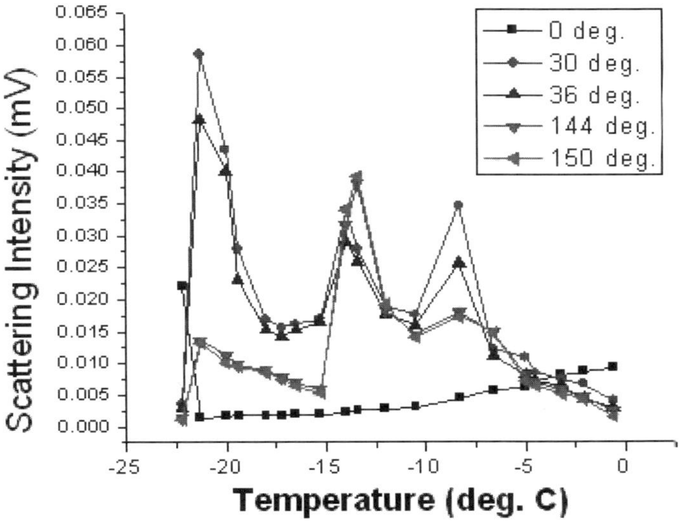

Fig. 5. Variation of scattering intensity at different angles against seeding temperature.

and 36° are of the order of mJ/m, whereas, in general, the corresponding intensities at 144° and 150° are one or two orders less. The temperature variation of maximum scattered intensity at different seeding angles is presented in Fig. 5. All these values are the mean of four observations. The direct intensity sharply falls as the seeding temperature increases from −22°C to −21°C, and then it shows a slow rise as the temperature rises. This is understandable, as below −21°C, the crystals are quite insignificant in number as well as size. As the crystal number and modal dimension show an overall gradual fall with a rise in temperature beyond −21°C, the direct intensity also slowly rises.

In case of scattered intensity, there are three distinct peaks in forward-scattering intensity (at 30° and 36°) at −21°C, close to −14°C, and −8°C. The peaks at −21°C and close to −14°C are understandable, as both the crystal number and modal dimension show peak at those temperatures. The highest peak naturally occurs at −21°C. But, the justification for the peak in forward-scattering intensity at −8°C is unclear.

In case of backward-scattering intensity (at 144° and 150°), the highest peak occurs at −13°C. The backward-scattering at −13°C is even greater than the forward-scattering intensity at the corresponding temperature. Nonspherical ice crystals are known to produce various features like corona and halo.[30–32] It is reported that hexagonal crystals can produce halo when back-scattering intensity significantly increases. This happens mainly due to successive reflection and refraction in a hexagonal crystal.[33] Since at the seeding temperature of −13°C, the crystals are dominantly hexagonal in shape and also there are some cubic and prism-shaped crystals, the dominant back-scattering intensity at −13°C seems to be due to halo effect. In fact, strong back-scattering effect is generally observed in cirrus cloud top region.[31]

The optical parameters like scattering coefficient, extinction coefficient, and optical thickness are three important optical parameters, characterizing a crystal cloud. The variation of scattering coefficient with seeding temperature is presented in Fig. 6, for different angles of scattering. The scattering coefficient is found to attain the peak value at −21°C for all angles of scattering. It shows a second peak close to −13°C. The back-scattering coefficient at 144° is found to be greater than the forward-scattering coefficient at 36° at that temperature. Around −18°C to −16°C, the back-scattering coefficient at 144° is found to have the maximum value, as cubic crystals are dominant. However, the absolute value is quite less there. A representative diagram for the variation of extinction coefficient

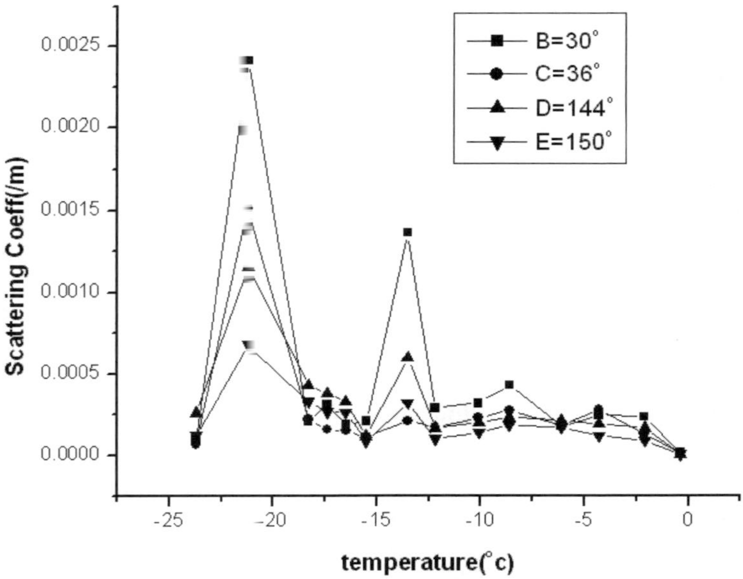

Fig. 6. Variation of scattering coefficient with seeding temperature.

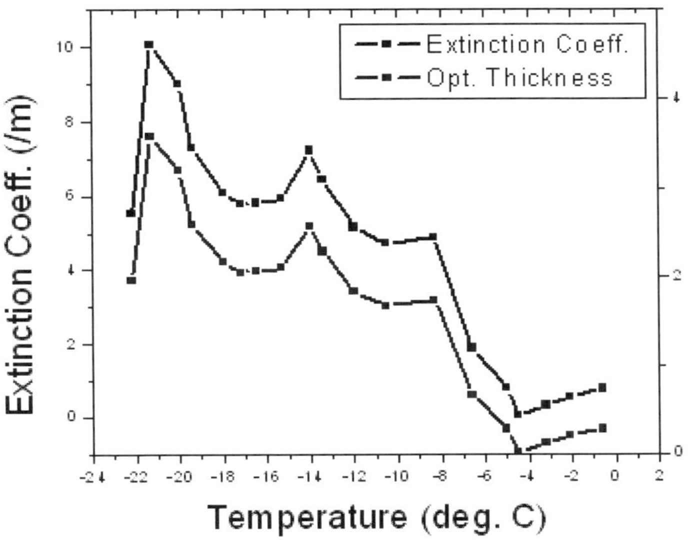

Fig. 7. Variation of extinction coefficient and optical thickness with seeding temperature.

and optical thickness with seeding temperature is presented in Fig. 7. These curves are prepared from the scattering intensities at 30°. As expected, the curves for both the parameters have three peaks at −21°C, close to −14°C, and −8°C. A plot for extinction coefficient and the ratio of scattering intensity at 30° to extinction coefficient against time counted from the instant of seeding is presented in Fig. 8. Time variation of extinction coefficient from the instant of seeding is also presented in the same figure. The curve becomes almost flat only 15 s after seeding. During the later period, the actual intensity shows some fluctuations though the above-mentioned ratio remains more flat. This result confirms an earlier observation made by Saunders *et al.*[34]

4.7. *Growth and decay of crystals*

The growth and decay of scattering intensity with time, and the corresponding fall and rise of direct intensity are presented in Fig. 9. This is only a representative curve, and only two scattering angles of 30° and 36° are presented. Time is again counted from the instant of seeding. The diagram clearly reveals growth and subsequent fall of the crystals. From the

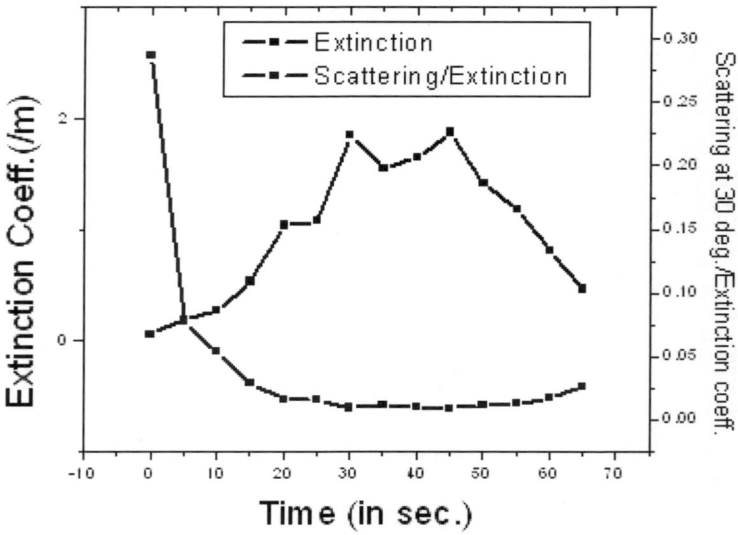

Fig. 8. The extinction coefficient and the ratio of scattering intensity at 30° to the extinction coefficient against time.

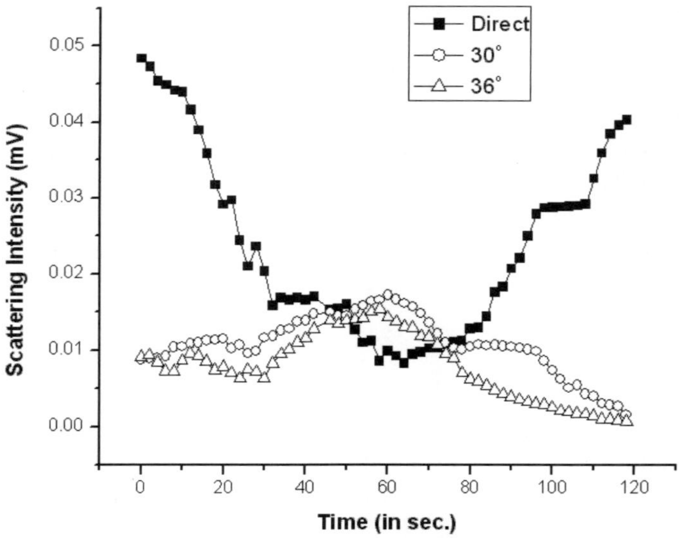

Fig. 9. The growth and decay of scattering intensity with time.

time variation of direct intensity, one can conclude that within 2 min after seeding all the crystals precipitate out. Crystal formation becomes complete within the first 30 s, thereafter the crystals remain floating for nearly 50 s, and falling down of crystals becomes complete in the next 40 s.

5. Conclusion

The widespread presence of DMS in atmosphere and ocean is now established. DMS has strong possibility of being oxidized in the atmosphere, and then DMSO may be a product. Reports are also available for the production of DMSO due to biosynthesis of DMS by phytoplanktons. Some DMSO may be lifted to middle and upper atmosphere by convection, and then it has a strong possibility of acting as IN. It is known that nearly 30% of the earth is always covered by cirrus cloud,[35–38] which is mainly formed by ice crystals, and DMSO is supposed to have some contribution to it. So, DMSO is expected to play a role in the modification of the solar energy budget of the earth.

It is noted that DMSO has good ice-nucleating ability over a long range of temperature on the negative side of 0°C. A sharp peak in that ability is attained at the eutectic temperature of DMSO and water combine. DMSO can exist in liquid state over the tropical ocean surface and it has then the chance of coating various particulate aerosols. It may be noted that DMSO-coated Ammonium Sulfate particles have much greater nucleation ability. The study of nucleation characteristics of other DMSO-coated solid aerosols is also needed to assess the extent of its influence in the solar energy modification on earth.

Forward-scattering of the heterogeneous ice cloud formed by DMSO is significant. Though the back-scattering is in general less than the forward-scattering by an order of 1 or 2, it attains a value more than the forward-scattering from −18°C to −13°C, when cubic and hexagonal crystals dominate. It seems, the halo effect then dominates. In the laboratory, when convection is absent, all crystals due to DMSO fall down within 2 min of seeding.

Acknowledgment

The authors' thanks are due to Indian Meteorological Department, Government of India for the sanction of the research project. The present work is a part of the project.

References

1. P. Warneck, *Chemistry of the Natural Atmosphere* (Academic Press, 2000).
2. P. L. Hanst, L. L. Spiller, D. M. Watts, J. W. Spence and M. F. Miller, Infrared measurements of fluorocarbons, carbon tetrachloride, carbonyl sulfide and other atmospheric trace gases, *J. Air Pollutant Control Assoc.* **25** (1975) 1220–1226.
3. F. J. Sandals and S. A. Penkett, Measurements of carbonyl sulfide and carbon disulfide in the atmosphere, *Atmos. Environ.* **11** (1997) 197–199.
4. P. J. Maroulis and A. R. Bandy, Estimate of the contribution of biologically produced dimethyl sulfide to the global sulfur cycle, *Science* **196** (1977) 647–648.
5. F. B. Hill, V. P. Aneja and R. M. Felder, A technique for measurements of biogenic sulfur emission fluxes, *J. Environ. Sci. Health* **A13** (1978) 199–225.
6. J. E. Lovelock, R. J. Maggs and R. A. Rasmussen, Atmospheric dimethyl sulfide and the natural sulfur cycle, *Nature* **237** (1972) 452–453.
7. B. C. Nguyen, A. Gaudry, B. Bonsang and G. Lambert, Reevaluation of the role of dimethyl sulfide and the natural sulfur cycle, *Nature* **237** (1978) 452–453.
8. W. R. Barnard, M. O. Andreae, W. E. Watkins, H. Bingemar and H. W. Georgii, The flux of dimethyl sulfide from the oceans to atmosphere, *J. Geophys. Res.* **87** (1982) 8787–8793.
9. I. V. Patroescu, I. Barnes, K. H. Becker and N. Mihalopoulos, FT-IR product study of the OH-initiated oxidation of DMS in the presence of NO_x. *Atmos. Environ.* **33** (1999) 25–35.
10. S. Sorensen, H. Falbe-Hansen, M. Mangoni, J. Hjorth and N. R. Jensen, Observations of DMSO and CHS(O)OH from the gas phase reaction between DMS and OH, *J. Atmos. Chem.* **24** (1996) 299–315.
11. I. Barnes, K. H. Becker and I. Patroescu, FTIR product study of OH initiated oxidation of dimethyl sulfide: Observation of carbonyl sulfide and dimethyl sulfoxide, *Atmos. Environ.* **30** (1996) 1805–1814.
12. C. Arsene, I. Barnes and K. H. Becker, New product and aerosol studies on the photo-oxidation of dimethyl sulfide, *Proc. US/German Environmental Chamber Workshop*, Riverside, California, USA (http://pah.cert.ucv.edu/~carter/epacham/meeting1.html).
13. L. Darroch, Dimethylsulphoxide (DMSO) in seawater, http://www.uea.ac.uk/env/marinegas/research/dmso.shtml (2005).
14. A. P. Pszenny, G. R. Harvey, C. J. Brown, R. F. Lang, W. C. Keene, J. N. Galloway and J. Merrill, Measurements of dimethyl sulfide oxidation products in the summertime North Atlantic marine boundary layer, *Global Biogeochem. Cycles* **4** (1990) 367–379.
15. R. J. Charlson, J. E. Lovelock, M. O. Andrae and S. G. Warren, Oceanic phytoplankton, atmospheric sulfur, cloud albedo and climate, *Nature* **326** (1987) 655–661.
16. M. Legrand, C. Feniet-Saigne, E. S. Saltzman, C. Germain, N. I. Barkov and V. N. Petrov, Ice-core record of oceanic emissions of dimethyl sulfide during the last climate cycle, *Nature* **350** (1991) 144–146.

17. H. Berresheim, F. L. Eisele, D. J. Tanner, D. S. Covert, L. McInnes and D. C. Ramsey-Bell, Atmospheric sulfur chemistry and cloud condensation nuclei (CCN) concentrations over the northern Pacific coast, *J. Geophys. Res.* **98** (1993) 12701–12711.
18. A. Hazra, S. Paul, U. K. De, S. Bhar and K. Goswami, Investigation of ice nucleation/hydrate crystallization by aqueous solution of ammonium sulfate, *Prog. Cryst. Growth Char. Mater.* **47** (2004) 45–61.
19. S. Paul, A. Hazra, D. Roy, U. K. De, S. Bhar and K. Goswami, Ice nucleating behaviour of aqueous and alcoholic solution of Phloroglucinol: A laboratory study, *J. Weather Mod.* **36** (2004) 41–46.
20. A. Hazra, S. Paul, U. K. De, S. Kar and K. Goswami, Study of ice nucleation/hydrate crystallization over urea, in *Trends in Crystal Growth Research*, ed. G. V. Karas (Nova Science Publications Inc., NY, 2005), ISBN: 1-59454-541-3, Chap. 9.
21. S. Paul, L. N. Biswas, U. K. De and K. Goswami, Nucleation and scattering properties of ice cloud due to seeding of sodium chloride as aqueous solution and dust, *Atmos. Environ.* **39** (2005) 6213–6222.
22. S. Paul, A. Hazra, U. K. De, S. Bhar and K. Goswami, Comparative study of nucleation by different alcoholic solutions of Benzoin and Benzoin dust, *J. Atmos. Chem.* **53** (2006) 155–168.
23. A. Hazra, U. K. De and K. Goswami, Nucleating characteristics of *Pseudomonas aeruginosa* above 0°C and the role of L-leucine, *J. Cryst. Growth* **286** (2006) 114–120.
24. P. D. Hien, V. T. Bac and N. T. H. Thinh, Investigation of sulfate and nitrate formation on mineral dust particles by receptor modeling, *Atmos. Environ.* **39** (2005) 7231–7239.
25. C. P. R. Saunders and N. A. Wahab, The replication of ice crystals, *J. Appl. Meteorol.* **12** 1035–1039.
26. J. Hallett, *Cloud Particle Replicator for Use on a Pressurized Aircraft: Part I*, Operating manual (Air Force Geophysics Laboratory, United States Air Force, Hanscom, AFB, MA, 1976).
27. L. V. Azaroff, *Introduction to Solids* (McGraw Hill, New York, 1995).
28. M. Hanajima, On the condition of growth of snow crystal, *Low Temp. Sci.* **A1** (1954) 53–65.
29. R. R. Rogers and M. K. You, *A Short Course in Cloud Physics*, 3rd Edn. (Paragon Press, Oxford, 1989), p. 156.
30. K. Sassen, Corona producing cirrus clouds properties derived from polarization lider and photographic analysis, *Appl. Optics* **30** (1991) 3421–3428.
31. K. Sassen, G. C. Mace, J. Hallett and M. R. Poellot, Corona-producing ice clouds: A case study of a cold mid-latitude cirrus layer, *Appl. Optics* **37** (1998) 1477–1485.
32. M. Vollmer and R. Tammer, Laboratory experiments in atmospheric optics, *Appl. Optics* **37** (1998) 1557–1568.
33. E. Trankle and R. G. Greenler, Multiple-scattering effects in halo phenomena, *J. Optical Soc. Am.* **A4** (1987) 591–599.

34. C. P. R. Saunders, J. Rimmer, P. Jonas, J. Arathon and C. Liu, Preliminary laboratory studies of the optical scattering properties of the crystal clouds, *Annales Geophysicae* **16** (1998) 618–627.
35. V. E. Ramsnathan, E. J. Pitcher, R. C. Malone and L. Blackman, The response of a spectral general circulation model to refinements in radiative process, *J. Geophys. Res.* **102**(D20) (1983) 23845–23850.
36. K. N. Liou, Influence of cirrus clouds and climate process: A global perspective, *Monthly Weather Rev.* **114** (1986) 1167–1199.
37. E. J. Jensen, O. B. Toon, D. L. Westpal, S. Kinne and A. J. Heymsfield, Microphysical modeling of cirrus: 1. comparison with 1986 FIREIFO measurements, *J. Geophys. Res.* **99**(D5) (1994) 10421–10442.
38. E. J. Jensen, O. B. Toon, D. L. Westpal, S. Kinne and A. J. Heymsfield, Microphysical modeling of cirrus: Sensitivity studies, *J. Geophys. Res.* **99**(D5) (1994) 10443–10454.

Advances in Geosciences
Vol. 10: Atmospheric Science (2007)
Eds. J. H. Oh and G. P. Singh

IMPACT ASSESSMENT OF GLOBAL TEMPERATURE PERTURBATIONS ON URBAN AND REGIONAL OZONE LEVELS IN SOUTH TEXAS

JHUMOOR BISWAS, KURUVILLA JOHN and ZUBER FAROOQUI
CREST-RESSACA, Frank H. Dotterweich College of Engineering, Texas A&M University-Kingsville, Kingsville, Texas 78363

The recent Intergovernmental Panel on Climate Change report predicts significant temperature increases over the century which constitutes the pulse of climate variability in a region. A modeling study was performed to identify the potential impact of temperature perturbations on tropospheric ozone concentrations in South Texas. A future case modeling scenario which incorporates appropriate emission reduction strategies without accounting for climatic inconsistencies was used in this study. The photochemical modeling was undertaken for a high ozone episode of 13–20 September 1999, and a future modeling scenario was projected for ozone episode days in 2007 utilizing the meteorological conditions prevalent in the base year. The temperatures were increased uniformly throughout the simulation domain and through the vertical layers by 2°C, 3°C, 4°C, 5°C, and 6°C, respectively in the future year modeling case. These temperature perturbations represented the outcome of extreme climate change within the study region. Significantly large changes in peak ozone concentrations were predicted by the photochemical model. For the 6°C temperature perturbation, the greatest amplification in the maximum 8-h ozone concentrations within urban areas of the modeling domain was approximately 12 ppb. In addition, transboundary flux from major industrialized urban areas played a major role in supplementing the high ozone concentrations during the perturbed temperature scenarios. The Unites States Environmental Protection Agency (USEPA) is currently proposing stricter 8-h ozone standards. The effect of temperature perturbations on ozone exceedances based on current and potential stringent future National Ambient Air Quality Standards (NAAQS) was also studied. Temperatures had an appreciable spatial impact on the 8-h ozone exceedances with a considerable increase in spatial area exceeding the NAAQS for the 8-h ozone levels within the study region for each successive augmentation in temperature. The number of exceedances of the 8-h ozone standard increased significantly with each degree rise of temperature with the problem becoming even more acute in light of stricter future proposed standards of ozone.

1. Introduction

The United States Environmental Protection Agency (USEPA) has prescribed the usage of certain types of air quality models to analyze characteristic ozone episodes and conceptualize ozone abatement strategies for ozone non-attainment regions in the corresponding State Implementation Plan (SIP).[1] Each SIP must demonstrate through computer modeling analyses and estimates that any suggested regulatory proposals will enable the pollution levels in the region to meet federal air quality standards in the future. The modeled future case helps determine a hypothetical future scenario using contemporary meteorology and potential emission reductions and enables the planners to determine whether the decision-making processes to reduce emissions are relevant in the context of future growth and development. However, the current decision-making process does not account for possible variations in ozone concentrations in the future due to potential changes in climate. The interaction between air quality and climate is an interactive process since resultant climate changes impact global atmospheric chemistry and background levels of air pollutant concentrations. The Intergovernmental Panel on Climate Change report by Meehl *et al.*,[2] based on results from most recent climatic models, predicts an average rise of global temperature between 1.4°C and 5.8°C by the year 2100. This significant rise in ambient temperature can impact global tropospheric chemistry as suggested by Fiore *et al.*,[3] and can therefore alter the chemical composition of the troposphere and affect both the surface ozone concentrations and ozone exceedances on regional and urban scales. Figure 1 highlights the key temperature-dependent photochemistry involved in the ozone formation.

Meteorological parameters which influence advection, dispersion, dilution, and rates of atmospheric chemical mechanisms affect air quality variabilities in most regions. Therefore, it is imperative to account for climate changes in numerical modeling experiments while developing emission control policies for the future. Several past studies, such as those by Seaman *et al.*,[4] and Sillman *et al.*,[5] have made detailed impact assessment of meteorological conditions on the surface ozone concentrations and ozone exceedance events. The study of ozone sensitivity to different modifications in temperatures employing process modeling and chemical transport models as described by Baertsch-Ritter *et al.*,[6] has revealed that increases in peak ozone concentrations are directly related to temperature increases and higher temperatures are usually associated with elevated

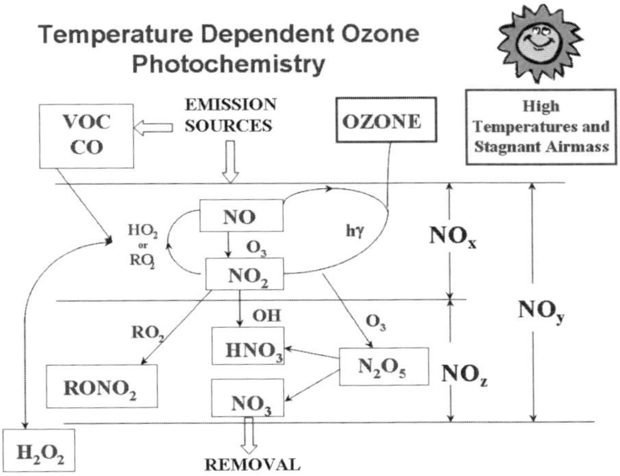

Fig. 1. Atmospheric photochemical pathways for the formation of ozone.

ozone concentrations. Temperature changes affect chemical dynamics and emission rates of anthropogenic and biogenic ozone precursors. It has also been established that temperature variations have the largest impact on peak ozone concentrations and ozone exceedances amongst all meteorological variables as per Dawson *et al.*[7] Therefore, it becomes important to assess the impact of future global temperature changes and consequently climate change on air quality decision-making processes.

The south Texas region has experienced several exceedances of the 8-h National Ambient Air Quality Standards (NAAQS) within different urban and semi-urban regions during the recent years. The region is characterized by a unique climatology of semi-arid coastal and inland areas and consists of a number of urban areas classified by the Texas Commission on Environmental Quality (TCEQ) to be in near non-attainment status of the 8-h ozone standards. These near non-attainment areas (NNA) have voluntarily opted to develop appropriate planning processes to continue to remain in attainment of the federal 8-h ozone standards. The future modeling scenario considered here was developed as a part of an attainment demonstration process. The emission estimations from anthropogenic sources were adjusted to account for such factors as anticipated growth or decline in population and economy as well as the impact of federal, state, and regional emission reduction measures.[8] The biogenic emissions inventory for the 2007 future case was the same as that used in the 1999 base case, and this was developed in accordance with the USEPA guidelines.[1]

There has been little focus so far on the effect of temperature perturbations on regional and urban ozone quality in the South Texas region. Emission control policies in these parts are currently implemented assuming that the climate conditions remain constant. This study illustrates the significant impact climate changes can have on potential ozone concentrations in the future, despite the assimilation of substantial emission reduction procedures. It examines the impact of an array of temperature perturbations on the model-simulated 8-h averaged ozone concentrations for the 2007 future case where emissions and other meteorological parameters were held constant. This study provides an in-depth analysis of the importance of the role of possible climate variability on surface ozone concentrations in South Texas by investigating spatial and temporal responses in modeled peak surface ozone concentrations and ozone exceedances to various temperature perturbations. The promulgation of a longer-term 8-h ozone standards since 1997 based on more rigorous health assessments by the USEPA resulted in an overall increase of non-attainment areas. Spatial and temporal scales are intrinsically linked in the ozone process as mentioned by Rao *et al.*[9] Therefore, the possible lowering of federal ozone standards to 70–75 ppb in the future[10] will considerably enhance this problem. This issue of climate change then becomes even more critical if the 8-h standards become more stringent in the near future, and this is highlighted in the study described here.

The primary objective of this ozone sensitivity study is to enhance the awareness of decision-makers regarding climate change impact on surface ozone concentrations so that future potential emission abatement strategies may be developed, factoring climate change in the decision-making process for the South Texas region.

2. Model Description

The Comprehensive Air quality Model with extensions (CAMx) version 3.1,[11] has been applied to assess the impact of perturbed temperatures in the South Texas region. The photochemical model CAMx that simulates various atmospheric physical and chemical processes was applied in a nested grid mode in this study. The base case modeling simulations were performed for high ozone days of 13–20 September 1999 that occurred in South Texas. The future case modeling simulations were conducted for 2007 projected to have a similar set of episodic days. Meteorological fields, boundary and initial conditions, dry deposition algorithms, chemical mechanisms, and other model configurations remained identical between September 1999 and 2007 simulations. The future case incorporated the 1999 base case meteorology, and the model was run under varying levels of temperature perturbations. The photochemical model grid system was in Lambert conformal projection system centered at $-100°$ and $40°$ with standard parallel latitudes at $30°$ and $60°$, respectively. The coarse outermost grid of the photochemical modeling domain with a grid resolution of 36 km covers south, southwest, and central parts of continental United States. The inner grid of 12 km grid resolution envelops the urban centers of Houston–Galveston area, Beaumont–Port Arthur area, Dallas–Fort Worth area, and the eastern Texas region. The innermost grid of 4 km grid resolution containing 90×108 cells is focused on south and central Texas and covers all the urban areas including Austin, San Antonio, Victoria, and Corpus Christi. The 12 vertical layers in CAMx extend from the surface all the way up to 4 km.

The Emission Processing System version 2.0,[12] was utilized to process emission inventory over the modeling domain and generate CAMx-ready input gridded emission files. A report by ENVIRON,[13] contains the details of the emissions inventory and emissions processing for the ozone episode of September 1999. Table 1 highlights the emission loads for each of the urban area in the south and central Texas.

Hourly meteorological data from 13–20 September 1999 were simulated using Fifth Generation Pennsylvania State University/National Center of Atmospheric Research (PSU/NCAR) Meteorological Model (MM5) version 4.3., as described by Grell *et al.*,[14] in a three-way nested grid mode to produce gridded three-dimensional meteorological inputs needed for the CAMx photochemical model. The meteorological model used analysis grid

Table 1. Emission loads (in tons) for urban areas of south and central Texas during 13–20 September 1999.

NNA subdomain	Emissions (in tons)			
	Anthropogenic		Biogenic	
	NOx	*VOC*	*NOx*	*VOC*
Austin	1143	758	182	1862
San Antonio	1052	568	166	1560
Corpus Christi	1146	1393	323	329
Victoria	301	145	69	185

nudging and the TCOON (Texas Coastal Ocean Observation Network) data for observation-based nudging along the coast. The MM5 model consists of 28 half-sigma levels in the vertical. Details regarding the meteorological model simulations have been presented in the report by Emery *et al.*[15]

3. Methods

The base case modeling simulations for the September ozone episodic period were evaluated using statistical metrics prescribed by USEPA,[1] for the daily 8-h averaged and the maximum 8-h ozone concentrations. A series of additional sensitivity simulations were performed in which the surface and upper air temperatures were perturbed from 1°C to 6°C in accordance with the anticipated global temperature changes as predicted in the IPCC report by Meehl *et al.*[2] The perturbations were applied over the entire nested modeling domain while maintaining the other meteorological variables constant. This analysis was conducted for the future case with the high ozone episode days of 16–19 September 1999 after allowing for a spin-up of 3 days. Only data from the surface grid cells of the 4 km modeling domain covering the South Texas region was used in the analysis. The main metrics used for comparison in the study include spatial variations of differences of 8-h ozone maxima, the 8-h ozone exceedances, the number of 8-h ozone exceedances, and the number of surface grid cells exceeding the average 8-h ozone standard in these regions.

4. Results and Discussion

4.1. *Base case model evaluation*

The model evaluation for the base case model run was undertaken using USEPA[1] statistical criteria. It was found that all the performance metrics were within the limits set by the EPA performance criteria. Further details of model evaluation and validation can be obtained from the model evaluation report by ENVIRON.[16] The model does underpredict ozone concentrations at an observational site in the coastal region of Corpus Christi but on the days of highest ozone concentrations it performs relatively well as shown in Fig. 2. The model behaves satisfactorily in the inland areas of San Antonio and Austin, and it meets the EPA-prescribed performance metrics. The future case modeling simulations were accomplished after base case validation to test the impact of emission control strategies.

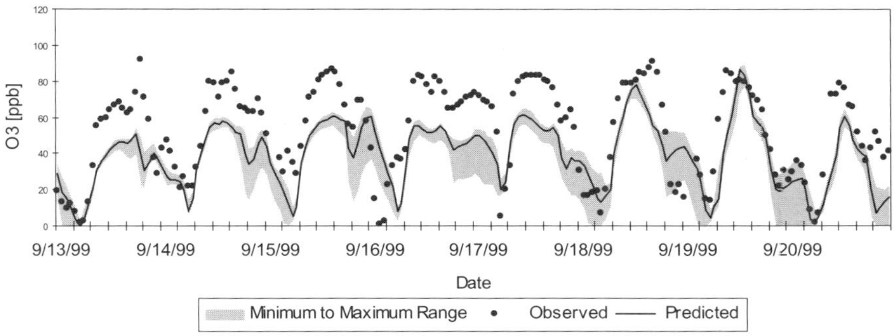

Fig. 2. Temporal representation of hourly ozone concentrations at a monitoring site in Corpus Christi, Texas.

4.2. *Spatial impact of perturbed temperature on peak ozone concentrations*

The spatial differences in emission distributions over the entire South Texas region have been summarized in Table 1. Although the peak ozone concentrations declined substantially in the South Texas region compared to the base case, the temperature perturbations radically impact future ozone concentrations. It was found that a 6°C perturbation produces the largest increase of ~12 ppb from the unperturbed case over the San Antonio region and the counties bordering Houston–Galveston area. Austin exhibited an increase of 9–12 ppb, as shown in Fig. 3. Corpus Christi revealed a maximum

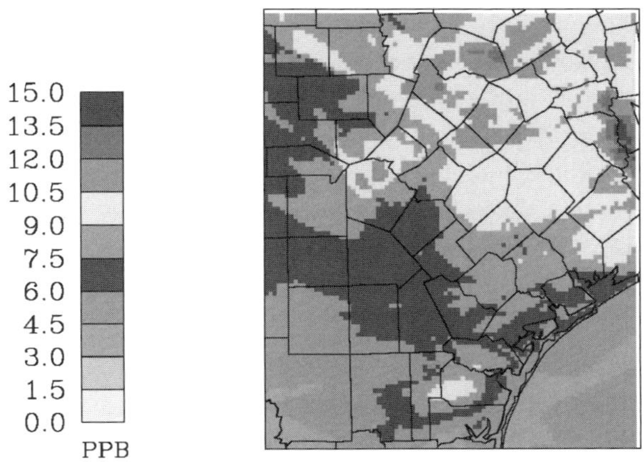

Fig. 3. Spatial distribution of the differences in episode maxima of 8-h ozone concentrations between the base case and the perturbed temperature (6°C) case.

Table 2. Maximum impact of perturbed temperature.

	+2°C	+3°C	+4°C	+5°C	+6°C
Difference in 8-h ozone conc. in ppb	6.98	8.71	10.13	11.45	12.59

impact of around 10 ppb. The spatial distribution pattern of the differences in the 8-h ozone episodic maxima between the base and the perturbed cases remains similar for all other perturbed temperature scenarios with the greatest impact occurring downwind of urban regions and particularly downwind of Houston–Galveston area. The utmost impact on the peak 8-h ozone concentrations is an increment of ~1.5 ppb from the preceding perturbed temperature case, as shown in Table 2 with the differences getting smaller for higher temperature perturbations.

4.3. *Peak ozone impacts at air quality monitoring sites in south texas*

The model-predicted 8-h averaged ozone concentrations were extracted at the Continuous Air Monitoring Sites (CAMS) maintained by TCEQ and located in Austin, San Antonio, Victoria, and Corpus Christi. Table 3 illustrates the impact of temperature perturbations on the model-predicted 8-h episodic peak ozone concentrations at each of the CAMS sites for each NNA subdomain. The 2007 future case exhibits a decline in the peak ozone concentrations of 5–6 ppb at the monitoring sites from the 1999 base case. However, with successive temperature perturbations, the peak 8-h ozone amplifications rise rapidly at the monitoring sites and exceed the base case ozone concentrations despite curbing of emissions, as shown in Table 3. This illustrates the importance of accounting for climate change impact while devising emission control strategies. The San Antonio area exhibits the highest episodic 8-h ozone concentrations amongst all the CAMS sites followed by Austin, Victoria, and Corpus Christi. The rate of increase of peak ozone values declines after the initial perturbation at these sites.

4.4. *Spatial extent of ozone exceedances*

Ozone exceedance values are important from the standpoint of health and hence from a regulatory perspective. Figures 4 and 5 characterize the spatial pattern of 8-h ozone exceedances greater than 84 ppb and 74 ppb for a

Table 3. Model-predicted maximum ozone concentrations (ppb) at South Texas urban sites.

NNA sub domain	CAMS	Perturbed temperatures						
		Base case	Future case	+2°C	+3°C	+4°C	+5°C	+6°C
Austin	03	81.36	76.41	81.41	82.86	84.1	85.19	86.17
	38	82.07	77.12	82.62	84.36	85.88	87.24	88.51
San Antonio	23	88.86	81.6	87.15	88.77	90.2	91.51	92.71
	58	94.43	85.52	91.04	92.57	93.91	95.11	96.22
	59	70.06	64.58	68.86	69.66	70.3	70.85	71.33
Corpus Christi	04	71.82	67.21	70.7	71.83	72.85	73.75	74.55
	21	76.79	70.84	74.77	76.11	77.3	78.37	79.34
Victoria	87	76.29	71.98	76.34	77.35	78.21	78.96	79.63

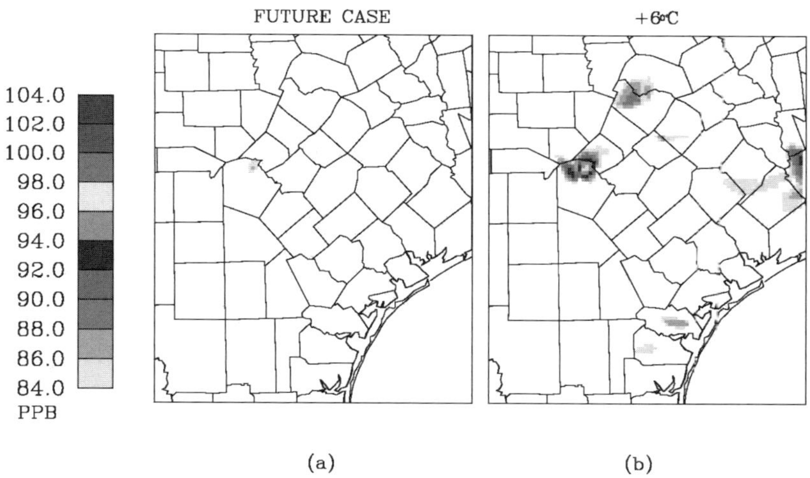

Fig. 4. Spatial pattern of ozone exceedances >84 ppb.

6°C temperature perturbation, which was chosen as a representative case for the future scenario. The maximum impact is noted downwind of the San Antonio region (~94 ppb), and is shown in Fig. 4. This is followed by northwest Austin and the counties north-east of Victoria, which are affected by transboundary pollution from Houston. Corpus Christi is less affected compared to the inland urban regions due to sea-breeze influence on air pollutants. In the case of the extreme 6°C perturbed scenario, the

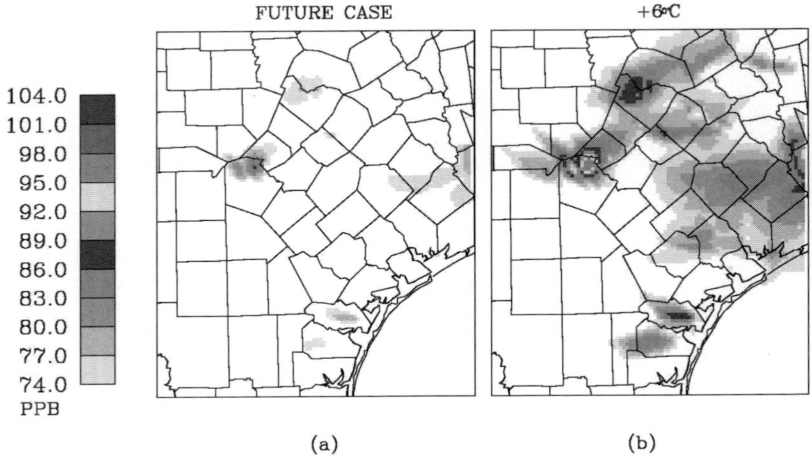

Fig. 5. Spatial pattern of ozone exceedances >74 ppb.

ozone exceedances become considerably more pronounced and widespread for levels greater than 74 ppb throughout the NNA domain, as shown in Fig. 5. The magnitude of surface ozone concentrations is once again highest in the San Antonio region, although ozone exceedances tend to be more prevalent in Austin, as shown in Table 5. The dependence of ozone formation on its precursors is nonlinear as described by Lin *et al.*,[17] and hence the ozone formation in each urban region is dependent on the VOC/NOx concentration ratios in the atmosphere. The ozone formation and destruction can occur in regions far away from their sources of emissions.

Furthermore, an analysis of the number of grid cells or area impacted by the 8-h ozone exceedances with rising temperature perturbations was undertaken. As shown in Table 5, for the first 2°C temperature augmentation there is a substantial increase of area with exceedances relative to the base case for the major portion of the 4 km domain. Thereafter, with each degree rise in temperature, the percent increase in area with 8-h ozone exceedances relative to the immediately preceding perturbation case declines, reinforcing the fact that certain portions of the modeling domain are more affected than others by temperature perturbations. As expected, the spatial extent of ozone exceedances increases significantly in comparison to the base case with rising temperatures as shown in Tables 4 and 5.

Table 4 portrays the fact that amongst the NNA regions, the San Antonio counties are the only area impacted by the 8-h ozone exceedances

Table 4. Number of grid cells exceeding 84 ppb.

	Perturbed temperature					
	Future case	+2°C	+3°C	+4°C	+5°C	+6°C
Austin	0	0	10	39	63	97
San Antonio	10	44	60	74	88	103
Corpus Christi	0	0	1	12	19	35
Victoria	0	0	0	0	0	0
Entire domain	10	63	98	161	247	409

Table 5. Number of grid cells exceeding 74 ppb.

	Perturbed temperature					
	Future case	+2°C	+3°C	+4°C	+5°C	−6°C
Austin	76	331	480	564	630	687
San Antonio	113	195	210	225	236	257
Corpus Christi	48	109	132	149	165	182
Victoria	0	37	93	126	134	146
Entire domain	430	1570	2220	2751	3150	3466

in the future case. Austin, Corpus Christi, and Victoria NNA regions have no instances of ozone exceedances in this case. However, Austin and Corpus Christi regions reveal a significant increase in the number of grid cells with ozone exceedances for the 4°C–6°C temperature perturbation cases relative to the unperturbed case. Overall, the inland areas show a much greater impact with increasing temperature perturbations than the coastal regions which is also consistent with the base case results.

Table 5 shows that in the event the standards are lowered in the future to 74 ppb, and that the pattern of potential spatial representation of exceedances in the future also changes. Austin will now have a greater spatial extent of exceedance values compared to San Antonio with increasing temperature perturbations. In the unperturbed case, although San Antonio still possesses the largest number of grid cells exceeding the standards with rising temperature perturbations, Austin has the largest percent of area affected by ozone exceedances. Lefohn *et al.*[18] analyzed that implementation of controls triggers fastest deterioration of ozone concentrations in those sites with highest daily maximum 8-h averaged concentrations relative to other sites. San Antonio seems to have benefited the most in the future scenario relatively. With the increasing temperature perturbations the ozone exceedances in the event of a lowered standard is not as extensive in

San Antonio as in Austin. Corpus Christi and Victoria also show substantial impact in this case in comparison to the case where ozone is greater than 84 ppb. This illustrates the increasing need of considering climate change scenarios while evaluating the impact of a tighter ozone standard.

4.5. *Number of 8-h ozone exceedances*

The impact of temperature perturbations on the surface ozone concentrations was also determined by computing the total number of 8-h averaged ozone exceedances over the entire 4 km modeling domain over the episodic period for each of the temperature perturbation scenarios. Although the number of exceedances has decreased substantially in the future case in comparison to the base case, with rising temperatures, the number of exceedances shows a significant increase. The first 2°C rise in temperature causes a large increase in the 4 km modeling domain for both the cutoff values (>84 ppb and >74 ppb), as shown in Tables 6 and 7. Subsequently, the increase in the number of 8-h ozone exceedance events is more moderate with respect to the immediately preceding temperature perturbation scenario. This analysis emphasizes the fact that the temperature perturbations affect

Table 6. Number of ozone exceedances (>84 ppb).

	Perturbed temperature					
	Future case	+2°C	+3°C	+4°C	+5°C	+6°C
Austin	0	0	19	68	163	263
San Antonio	18	125	183	254	302	368
Corpus Christi	0	0	1	20	36	69
Victoria	0	0	0	0	0	0
Entire domain	18	171	282	462	711	1124

Table 7. Number of ozone exceedances (>74 ppb).

	Perturbed temperature					
	Future case	+2°C	+3°C	+4°C	+5°C	+6°C
Austin	178	1406	2235	3076	3934	4858
San Antonio	413	1007	1186	1384	1594	1852
Corpus Christi	115	362	463	566	653	741
Victoria	0	131	291	482	631	786
Entire domain	1171	6690	10348	14441	18685	23035

the number of exceedances in certain grid cells in the modeling domain located mostly within the urban regions.

As shown in Table 6, San Antonio NNA region is the most affected area amongst the NNAs with 18 8-h ozone exceedances in the base case. Corpus Christi and Austin do not possess any 8-h ozone exceedances in the base case, but the number of ozone exceedances in these areas increases significantly with every 1°C rise in temperature above 3°C. In the case of San Antonio, the consequent increase in the number of exceedances with the rise of temperature is not as substantial as that observed in Austin especially for exceedances greater than 74 ppb. This analysis also indicates that ozone exceedances greater than 74 ppb occur in both large and small urban regions resulting in considerable increase in the number of ozone exceedances throughout the study region in Texas.

5. Conclusion

An assortment of ozone sensitivity runs were completed with different perturbations in temperatures over south Texas for a future year simulation using the high ozone days of September 1999. The results categorically revealed that global temperature augmentations could significantly impact peak 8-h ozone concentrations and 8-h ozone exceedances especially in the urban regions of south Texas. Climate interactions play an increasingly important role in local and regional air quality background concentrations. Transboundary flux enhancements also supplement ozone concentrations in the perturbed temperature scenarios. These results prove conclusively that potential emission control strategies need to incorporate climate change in their formulation. Climate change can amplify pollution-related health effects due to increase in the number of ozone exceedances. This issue of climate change becomes even more crucial if the 8-h standards become more stringent in the future. The lowering of ozone standards will substantially enhance the number of instances of ozone exceedances as well as the area under non-attainment of the 8-h ozone standards with increased temperature perturbations.

Acknowledgments

This material is based upon the work supported by the Center for Research Excellence in Science and Technology — Research on Environmental Sustainability of Semiarid coastal arid (CREST-RESSACA) at Texas

A&M University, Kingsville through funding from the National Science Foundation (NSF) under Cooperative Agreement No. HRD-0206259. Any opinions, findings, and conclusions or recommendations expressed in this material are those of the author and do not necessarily reflect the views of the National Science Foundation. The author would also like to acknowledge the editorial support of Ms Mugdha Nayak in the preparation of this chapter.

References

1. EPA, Draft report on the use of models and other analyses in attainment for the 8-h ozone NAAQS, EPA-44/R-99-0004 (1999), pp. 157.
2. G. A. Meehl, T. F. Stocker, W. D. Collins, P. Friedlingstein, A. T. Gaye, J. M. Gregory, A. Kitoh, R. Knutti, J. M. Murphy, A. Noda, S. C. B. Raper, I. G. Watterson, A. J. Weaver and Z.-C. Zhao, Global climate projections, in *Climate Change 2007: The Physical Science Basis*, Contribution of Working Group I to the Fourth Assessment Report of the Intergovernmental Panel on Climate Change, eds. S. Solomon, D. Qin, M. Manning, Z. Chen, M. Marquis, K. B. Averyt, M. Tignor and H. L. Miller (Cambridge University Press, Cambridge, United Kingdom and New York, NY, USA, 2007).
3. A. M. Fiore, D. J. Jacob, B. D. Field, D. G. Streets, S. D. Fernandes and C. Jang, *Geophys. Res. Lett.* **29**(19) (2002) 1919.
4. N. L. Seaman and S. A. Michelson, *J. Appl. Meteorol.* **39** (2000) 384–398.
5. S. Sillman and P. J. Samson, *J. Geophys. Res.* **100** (1995) 14175–14188.
6. N. Baertsch-Ritter, J. Keller, J. Dommen and A. S. H. Prevot, *Atmos. Chem. Phys.* **4** (2004) 423–438.
7. J. P. Dawson, P. J. Adams and S. N. Pandis, *Atmos. Environ.* **41** (2006) 1494–1511.
8. U. T. Austin and ENVIRON, Report submitted to TCEQ, available online at http://www.tceq.state.tx.us/assets/public/implementation/air/sip/sipdocs/2004-06-AUS/AppendixI.pdf (2004).
9. S. T. Rao, I. G. Zurbenko, R. Neagu, P. S. Porter, J. Y. Ku and R. F. Henry, *Bull. Am. Meteorol. Soc.* **78** (1997) 2153–2166.
10. EPA, Review of the National Ambient Air Quality Standards for Ozone: Policy Assessment of Scientific & Technical Information (2006).
11. ENVIRON International Corporation, *User's Guide Comprehensive Air Quality Model with Extensions (CAMx)*, Version3.10 (2002).
12. EPA, *Volume IV: Users Manual for the Emissions Preprocessor System 2.0*: EPA-450/4-9007D(R): Office of Air Quality (1992).
13. ENVIRON International Corporation, Appendix D: 1999 Base Case Modeling Emissions Inventory Development, Report submitted to TCEQ, available online at http://www.tceq.state.tx.us/assets/public/implementation/air/sip/sipdocs/2004-06-SA/033104_AppD_1999EI_v2.pdf (2004).

14. G. A. Grell, J. Dudhia and D. R. Stauffer, NCAR Technical Note, NCAR TN-398-STR (1994), pp. 138.
15. C. A. Emery, E. Tai and G. McGaughey, Report submitted to TCEQ, available online at http://www.tceq.state.tx.us/assets/public/implementation/air/sip/sipdocs/2004-06-SA/04085sipapb_pro.pdf (2003).
16. AACOG, Appendix E: 1999 Base Case Development and Performance Analysis, Report submitted to TCEQ available online at http://www.tceq.state.tx.us/assets/public/implementation/air/sip/sipdocs/2004-06-SA/04085sipape_pro.pdf.
17. X. Lin, M. Trainer and S. C. Liu, *J. Geophys. Res.* **93** (1988) 5879–5888.
18. A. S. Lefohn, D. S. Shadwick and S. D. Ziman, *Environ. Sci. Technol.* **32**(11) (1998) 276A–282A.